TEACHING WITH

Volume 2

Edited by

Evan M. Maletsky
Montclair State College
Upper Montclair, New Jersey

NATIONAL COUNCIL OF TEACHERS OF MATHEMATICS

Copyright ©1993 by
THE NATIONAL COUNCIL OF TEACHERS OF MATHEMATICS, INC.
1906 Association Drive, Reston, Virginia 22091-1593
All rights reserved

ISBN 0-87353-369-0

The publications of the National Council of Teachers of Mathematics present a variety of viewpoints. The views expressed or implied in the publication, unless otherwise noted, should not be interpreted as official positions of the Council.

Printed in the United States of America

CONTENTS

Introduction iv

Golden Rectangles and Ratios 1
    Rick Billstein and Johnny W. Lott

Primes 7
    Margaret Kenney

Models: The Mathematician's Laboratory Experiment 13
    Glenn Allinger and Lyle Andersen

Polygonal Numbers 19
    Boyd Henry

Fair Games 27
    John Van Beynen

Geometry in a Circle 33
    Melfried Olson and Judy Olson

Paper-punching Patterns 39
    John Firkins

Midpoint Madness 45
    Lee Yunker, Dan Dolan, and John Van Beynen

Minimum Distance 51
    Jerry Johnson

Statistical Decision Making 57
    Gail Burrill

Directed Graphs 63
    Jack Burrill and Gail Burrill

Investigating Perimeter and Area 69
    Judy Mumme

Star Rec 75
    Maurice Burke

What Is a Curve of Constant Width? 81
    Millie Johnson

Ciphers 87
    Anne Teppo

Quadri-gon(e)? 95
    Gerald White, Melfried Olson, and Tanya Olson

Godzilla: Fact or Fiction? 101
    Rick Billstein and Jim Trudnowski

5-con Triangles 107
    Maurice Burke

Striking Sequences 113
    Carole B. Lacampagne

Invariants 119
    Melfried Olson and Glenn Bruckhart

INTRODUCTION

The *NCTM Student Math Notes* is a publication for students in middle, junior, and senior high school. It was begun in 1982 and appears five times a year with the *News Bulletin*. Each four-page issue focuses on a single theme, developed from a readily accessible opening page through some challenging extensions. The *Notes* are written for a wide audience of varied ages, backgrounds, and abilities.

The original intent of the *Student Math Notes* was to put in the hands of the mathematics students a publication that was designed—

- to stimulate and promote interest and enjoyment in mathematics;
- to illustrate mathematics as an exciting and challenging field;
- to supply material with which students could personally identify.

It soon became apparent that teachers saw a wider application for the *Notes*, including selected use in their own classrooms as concept openers and extensions and problem-solving and enrichment experiences. They also became a source of material for teacher training and in-service workshops. Today, these *Notes* clearly fit within the framework of the recommendations found in the NCTM *Curriculum and Evaluation Standards*, especially in enhancing experiences in—

- mathematics as problem solving;
- mathematics as communication;
- mathematics as reasoning;
- mathematical connections.

In 1987, the first compilation of the *Notes* appeared, bound in a single volume under the title *Teaching with Student Math Notes*. This publication extends the collection into a second volume. New four-page issues have been reproduced, along with teacher notes, detailed solutions, suggested extensions, and additional worksheets. As editor of both these publications and the first four years of the *Notes* themselves, I would like to extend my thanks to the many teachers, readers, and students who submitted manuscripts, contributed suggestions, identified mistakes, and offered support in these projects. My special thanks go to Lee Yunker, Dan Dolan, and Johnny Lott, who served as editors over the four years of the publications in this volume.

We are living in exciting, challenging, and changing times in mathematics education. All students need a broad encompassing view of mathematics, an understanding and appreciation of its power and beauty, and a wide variety of successful personal experiences with the subject. It is hoped that this NCTM publication can help in reaching those important goals.

EVAN M. MALETSKY

NATIONAL COUNCIL OF TEACHERS OF MATHEMATICS
STUDENT math notes
SEPTEMBER 1986

Golden Rectangles and Ratios

All the figures below have a rectangular shape. What else do they have in common? This particular rectangular shape, called the *golden rectangle,* is considered to be the most pleasing to the eye.

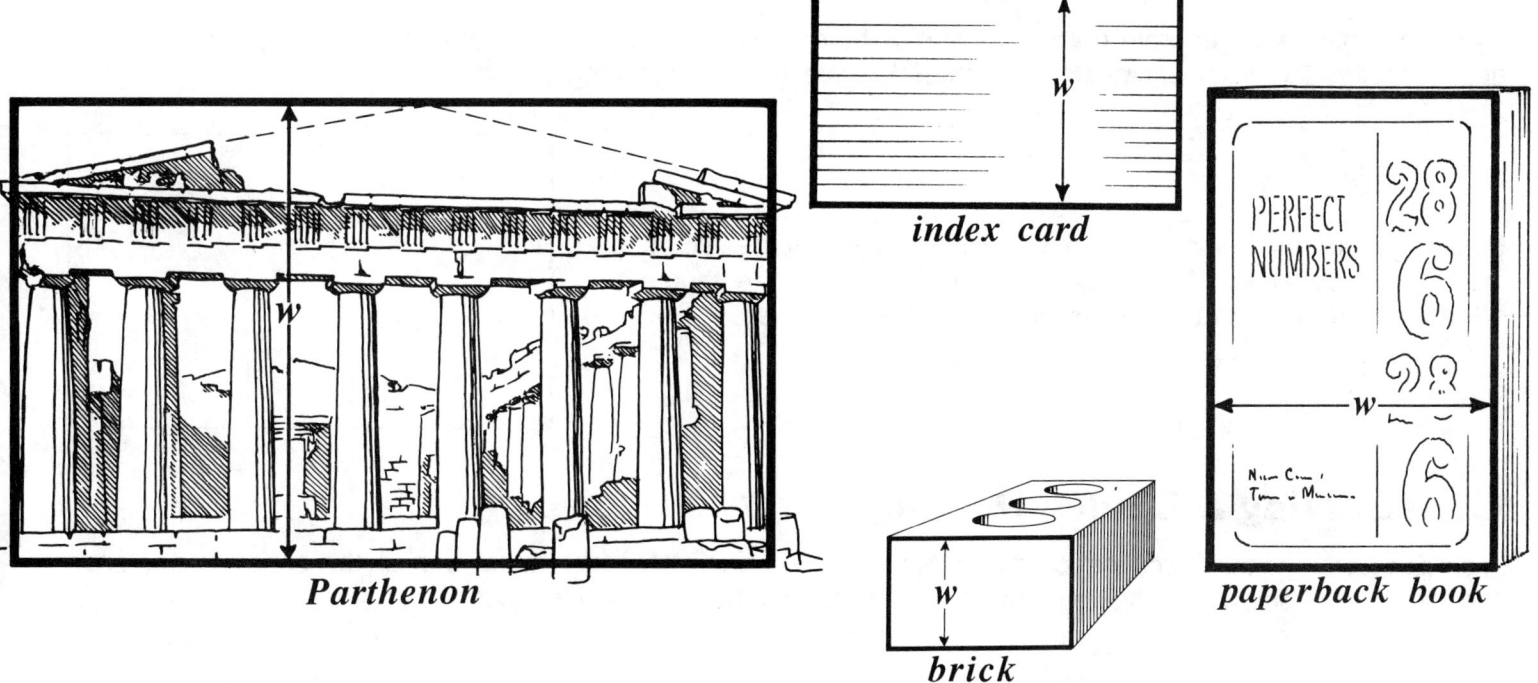

Measure the length and width of each rectangle in millimeters. Express the ratio of the width to the length as a decimal rounded to two places.

	Width (*w*)	Length (*l*)	*w*/*l*	Decimal Ratio
Parthenon	_____	_____	_____	_____
Index card	_____	_____	_____	_____
Paperback book	_____	_____	_____	_____
Brick	_____	_____	_____	_____

By expressing these ratios in decimal form, we can observe that each of them is approximately 0.6. The ratio 0.61803 . . . is called the *golden ratio*.

The editors wish to thank Rick Billstein and Johnny W. Lott, University of Montana, Missoula, MT 59812, for writing this issue of the *NCTM Student Math Notes.*

A Physical Model for Generating Golden Rectangles

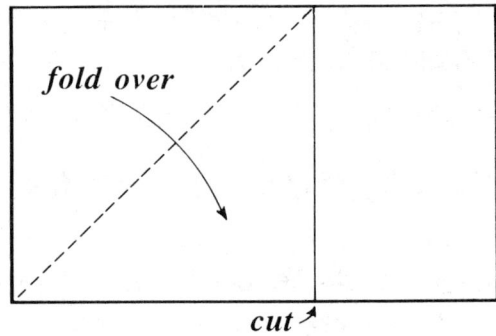

Cut a sheet of paper to measure 25 cm by 15.5 cm. This rectangle closely approximates a golden rectangle. Fold over one corner of the rectangle as shown in the adjacent figure. Then cut off the square from the rectangle. The remaining rectangle has the same proportions as the original rectangle; hence it is also a golden rectangle.

You can continue to generate golden rectangles by repeating this process. Each successive square has sides approximately 0.61803 ... times the length of the sides of the preceding square.

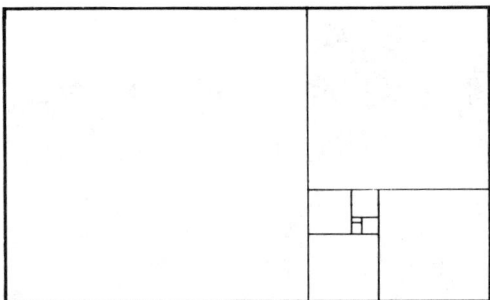

Place the squares together to form the original golden rectangle as in the figure shown here. This representation of the golden rectangle is often referred to as "the rectangle with the whirling squares."

Constructing a Golden Rectangle

A golden rectangle can be constructed using a compass and straightedge. To accomplish this construction, follow the steps given below. The figure shows the appropriate lettering of the vertices.

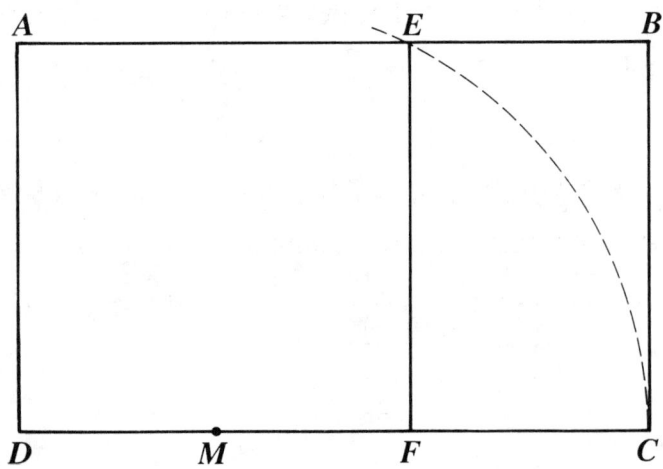

1. Construct a square AEFD.
2. Bisect $\overline{DF}$. Label the midpoint M.
3. Extend $\overline{DF}$.
4. With center M and radius ME, draw an arc intersecting $\overleftrightarrow{DF}$ at C.
5. Construct a perpendicular to $\overline{DC}$ at C.
6. Extend $\overline{AE}$ to intersect the perpendicular at B.

The rectangle ABCD is a golden rectangle. It can be shown that BCFE is also a golden rectangle. To show that this statement is true, assume MF = 1 and find each of the following lengths:

(a) FE _____ (b) BC _____ (c) ME _____ (d) MC _____ (e) FC _____ (f) DC _____

Find the decimal values of these ratios: BC/DC _____   FC/BC _____   Are they equivalent?

2 Teaching with *Student Math Notes*: Volume 2

Generating Golden Ratios via Logo

To generate a Logo procedure that draws successive golden rectangles using the idea of whirling squares, use Logo recursion. Recursion is the process of a procedure calling on a copy of itself. In mathematics, recursion can be thought of as a function that defines itself in terms of preceding terms.

First, write a procedure to draw a square of variable size that returns the turtle to its original position and heading.

```
TO SQUARE :SIZE
    REPEAT 4[FD :SIZE RT 90]
END
```

Next, add another square onto the square. However, you must first move to the upper-right-hand corner and give the turtle a heading of 90. You can do so by redrawing the first two sides of the square. Next, draw a square whose side is 0.61803 times the length of the original square. To generate the whirling squares, use recursion to repeat the process until you obtain a square that is as small as desired. This new procedure, used with the one above, will generate golden rectangles using whirling squares. (In Apple Logo, replace STOP with [STOP].)

```
TO GOLDEN.RECTANGLES :SIZE
    IF :SIZE < 1 STOP
    SQUARE :SIZE
    FD :SIZE RT 90 FD :SIZE
    GOLDEN.RECTANGLES :SIZE * 0.61803
    HIDETURTLE
END
```

An alternate procedure, called GOLDEN, for generating golden rectangles is given below. Note that it doesn't require the existence of the procedure SQUARE. Study it to determine how it works.

```
TO GOLDEN :SIZE
    IF :SIZE < 1 STOP
    REPEAT 2[FD :SIZE RT 90 FD :SIZE * 1.61803 RT 90]
    FD :SIZE RT 90 FD :SIZE
    GOLDEN * :SIZE * 0.61803
    HIDETURTLE
END
```

In the GOLDEN procedure, the first line names the procedure, the second line tells the procedure when to stop, and the third line draws the initial golden rectangle.

- What does the fourth line do?
- What does the fifth line do?

Draw the picture that would result if the fourth line were removed from the procedure.

Imagine drawing a 90-degree arc connecting the opposite corners of each square as shown here.

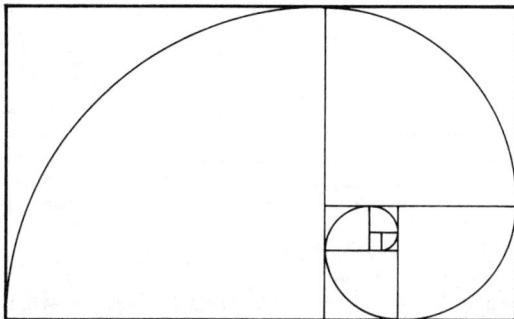

The resulting figure is called a *golden spiral*.

Teaching with *Student Math Notes:* Volume 2

The Divine Proportion

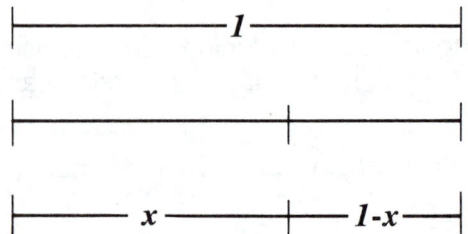

The divine proportion was derived by the fifteenth-century mathematician Luca Pacioli. It is found by dividing a segment into two parts so that the length of the smaller part is to the length of the larger part as the length of the larger part is to the length of the entire segment. Divide a segment one unit long into two parts; label the longer segment x, as shown.

The segments are in the divine proportion if the following is true: $\frac{1-x}{x} = \frac{x}{1}$

Solve for x:
$$x^2 = 1 - x$$
$$x^2 + x - 1 = 0$$
$$x = \frac{-1 + \sqrt{5}}{2} \text{ or } \frac{-1 - \sqrt{5}}{2}$$

Since lengths cannot be negative, the negative root can be discarded. Use a calculator to evaluate the positive root to five or more decimal places. Are you surprised at the result? (It's our new friend, the golden ratio, 0.61803 . . . .) The algebraic expression allows us to compute the value of the golden ratio more precisely than the paper-folding geometry used earlier.

Did You Know That . . .

- in his art, Michelangelo used the golden ratio?
- the factorial function is an example of recursion?
- the golden ratio is the limit of the ratio of successive terms in the Fibonacci sequence, which is related to such natural phenomena as pinecones, sunflowers, pineapples, and seashells?
- the reciprocal of 0.61803 . . . equals 1.61803 . . . ? Does any other number differ from its reciprocal by 1?

Can You . . .

- write a Logo procedure to draw a golden spiral?
- use Euclidean tools to construct a regular pentagon?
- find the ratio of the measure of the side of a regular pentagon to the measure of its diagonal?
- show that BC/AB is a golden ratio in triangle ABC?

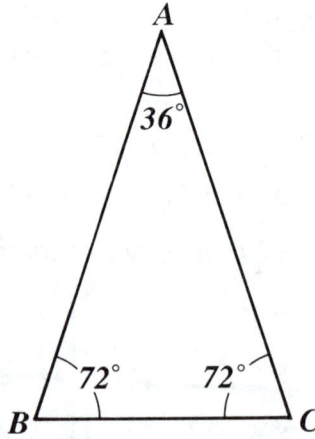

Teacher Notes

Page 1. The opening activity requires careful measuring in millimeters. The answers given below are the outside dimensions of the black borders of the rectangles.

Page 2. The hands-on activity produces a set of similar rectangles cut from paper. Stacked one on top of the other with a common vertex, the opposite vertices should fall on a straight line. This is because the ratio of the width to the length is constant at about $15.5 \div 25 = 0.62$.

The compass-and-straightedge activity gives a more precise construction of a golden rectangle. The required lengths should be found algebraically using the Pythagorean relationship.

Page 3. Logo is ideally suited for illustrating recursive properties inherent in the activities of the previous two pages. Mathematically, recursion implies a function defined through preceding terms. The Fibonacci sequence can be expressed in this recursive form.

Fibonacci sequence: $\quad x_{n+1} = x_n + x_{n-1}$

In generating successively smaller golden rectangles, the Logo procedure repeatedly calls up a copy of itself, each time with the size reduced by a factor of 0.61803.

Page 4. Although the number 0.61803... was referred to in each of the previous pages, only here is it carefully derived algebraically through the divine proportion. The golden ratio (0.61803...) and its reciprocal (1.61803...) differ by exactly 1. Solving the equation below leads to the same quadratic, and hence the same two roots, as the divine proportion. No other number differs from its reciprocal by 1.

$$\frac{1}{x} - x = 1$$

Answers

Page 1.

	w	l	$w \div l$	Decimal ratio
Parthenon	62	101	$62 \div 101$	0.61
Index card	29	49	$29 \div 49$	0.59
Paperback book	38	63	$28 \div 63$	0.60
Brick	15	25	$15 \div 25$	0.60

Page 2. a) 2 b) 2 c) $\sqrt{5}$ d) $\sqrt{5}$ e) $\sqrt{5} - 1$ f) $\sqrt{5} + 1$
$BC/DC = 0.61803...$; $FC/BC = 0.61803...$; Yes

Page 3. In the procedure called GOLDEN, the fourth line draws two adjacent sides of the square. The fifth line recalls the procedure with a new side of 0.61803 times the length of the preceding one.

An interesting extension to consider is that of finding the total length of the golden spiral given that the initial square has sides of 1. The computation involves the sum of a converging, infinite geometric sequence.

Numerical approximation: $\quad (1/2)\pi[1 + 0.61803 + (0.61803)^2 + (0.61803)^3 + \cdots] = 4.11236$

Algebraic solution: $\quad (1/2)\pi[1 + (-1 + \sqrt{5}) + (-1 + \sqrt{5})^2 + (-1 + \sqrt{5})^3 + \cdots] = \pi(3 + \sqrt{5})$

Page 6 (Extension). These are the first fifteen terms in the Fibonacci sequence:

1 1 2 3 5 8 13 21 34 55 89 144 233 377 610

Notice in the table on the left below how the decimal ratio of successive terms oscillates back and forth around the exact irrational value of $(-1 + \sqrt{5}) \div 2$ for the golden ratio. Clearly, the sequence of Fibonacci ratios is converging to the golden ratio.

8	$21 \div 34$	0.61764
9	$34 \div 55$	0.61818
10	$55 \div 89$	0.61797
11	$89 \div 144$	0.61805
12	$144 \div 233$	0.61802
13	$233 \div 377$	0.61803
14	$377 \div 610$	0.61803

```
1 → A
1 → B
Lbl 1
A+B
Pause
A+B → B
B-A → A
Goto 1
```

The program for the graphing calculator, shown above on the right, generates decimals for ratios of successive terms in the Fibonacci sequence. After thirty iterations, all eleven digits on the display remain fixed. Students familiar with graphing calculators may be interested in programming them to draw the same figures generated by the Logo procedure GOLDEN.

Extension—THE GOLDEN RATIO, FIBONACCI, AND PENTAGONS

The golden ratio is closely connected to the Fibonacci sequence and to the construction of a regular pentagon.

1. Write the next seven terms in the Fibonacci sequence.

   1   1   2   3   5   8   13   21   ___ ___ ___ ___ ___ ___ ___

2. Continue taking ratios of successive terms in the Fibonacci sequence as begun in the table below.

3. Complete the table by finding the decimal value for each ratio, correct to five decimal places. Toward what limiting value does this ratio appear to converge?

1	1 ÷ 1	1.00000	8	___	___
2	1 ÷ 2	0.50000	9	___	___
3	2 ÷ 3	0.66666	10	___	___
4	3 ÷ 5	0.60000	11	___	___
5	5 ÷ 8	___	12	___	___
6	8 ÷ 13	___	13	___	___
7	13 ÷ 21	___	14	___	___

4. Complete this construction of a regular pentagon using a compass and protractor.

   a. Bisect radius *OA*. Label the midpoint *M*. Draw segment *MB*.
   b. Using *M* as center and *MB* as radius, swing an arc through radius *OC*. Label the intersection point *P*.
   c. Using *B* as center and *BP* as radius, swing an arc through the circle. Label the intersection point *Q*.

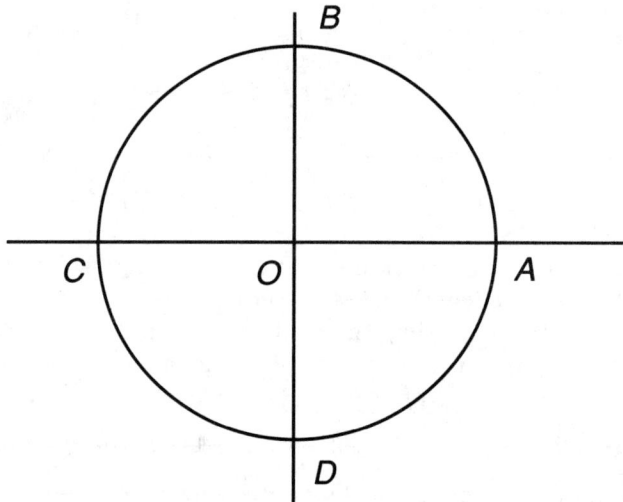

5. Segment *BQ* is the side of an inscribed pentagon. Mark off the remaining four sides around the circle using your compass. The ratio of the measure of a side of the regular pentagon to that of a diagonal is the golden ratio.

NATIONAL COUNCIL OF TEACHERS OF MATHEMATICS

Student Math Notes

NOVEMBER 1986

Primes

A *prime number* is a natural number that has exactly two factors, itself and 1. The pyramid below is called a *prime pyramid*. Each row in the pyramid begins with 1 and ends with the number that is the row number. In each row, the consecutive numbers from 1 to the row number are arranged so that the sum of any two adjacent numbers is a prime.

For example, look at row 5:

1) It must contain the numbers 1, 2, 3, 4, and 5.
2) It must begin with 1 and end with 5.
3) The sum of adjacent pairs must be a prime number.
4) $1 + 4 = \underline{5}$, $4 + 3 = \underline{7}$, $3 + 2 = \underline{5}$, and $2 + 5 = \underline{7}$.

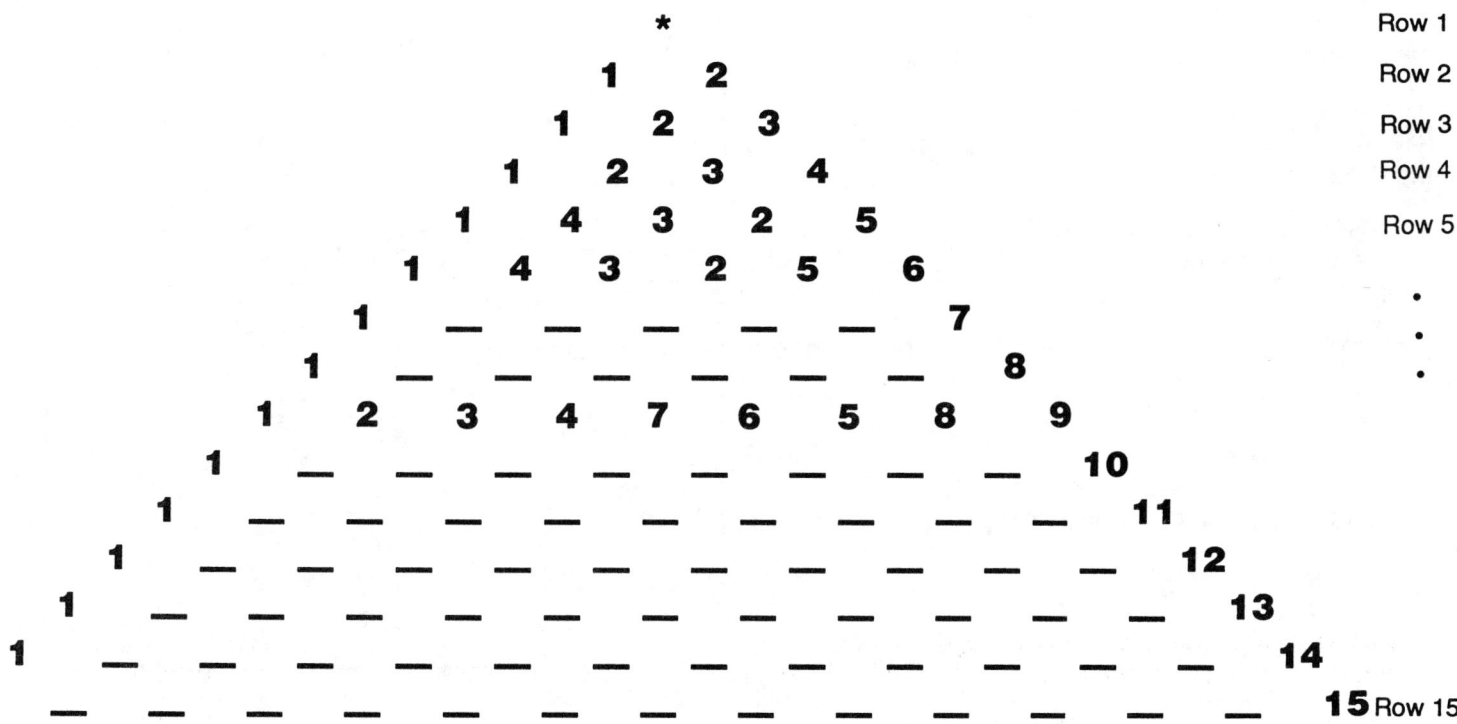

- Supply the missing numbers in this prime pyramid.
- Can you extend the prime pyramid beyond row 15?
- What patterns do you see in your solutions?
- What is your solution strategy for completing the pyramid?

The editors wish to thank Margaret Kenney, Mathematics Institute, Boston College, Chestnut Hill, MA 02167, for writing this issue of the *NCTM Student Math Notes*.

Prime Start

Euclid (born about 300 B.C.), a Greek mathematician famous for his geometry, showed that the number of primes is infinite. Eratosthenes (born in 230 B.C.), also a Greek, was the chief librarian at the University of Alexandria in Egypt. He developed a method for finding primes that is still in use today.

Sieve of Eratosthenes

Start with a 10 × 10 array of the numbers 1 through 100. The number 1 is crossed out because it is not a prime. Why not? _____

Circle 2, the first prime; then cross out the remaining multiples of 2 using a horizontal slash (—).

What visual pattern do the multiples of 2 make in the grid? _____

What's the next number that hasn't been crossed out? _____

Circle 3, the next prime; then cross out the remaining multiples of 3 using a vertical slash (|).

What visual pattern is made by the multiples of 3? _____

What is the next prime number that hasn't been crossed out? _____

Circle the 5 and cross out the remaining multiples of 5 using the diagonal slash (/).

In a similar manner, circle the next prime, 7, and cross out all its remaining multiples using the reverse diagonal slash (\ ).

~~1~~	2	3	4	5	6	7	8	9	10
11	12	13	14	15	16	17	18	19	20
21	22	23	24	25	26	27	28	29	30
31	32	33	34	35	36	37	38	39	40
41	42	43	44	45	46	47	48	49	50
51	52	53	54	55	56	57	58	59	60
61	62	63	64	65	66	67	68	69	70
71	72	73	74	75	76	77	78	79	80
81	82	83	84	85	86	87	88	89	90
91	92	93	94	95	96	97	98	99	100

At this point, except for 2, 3, 5, and 7, you have sifted out all the multiples of 2, 3, 5, and 7.

1. Are any multiples of 11, the next prime, not yet crossed out? _____

2. Why is this the case? _____

_____

The numbers that are not crossed out are the primes less than 100. Make a list of these primes.

2, 3, 5, 7, 11, ___, ___, ___, ___, ___, ___, ___, ___, ___, ___, ___, ___, ___, ___, ___, ___,

___, ___, ___, and ___.

Teaching with *Student Math Notes:* Volume 2

Twin Primes

Several pairs of primes in the list of primes less than 100 have a difference of 2. For example, the pairs 3 and 5, 5 and 7, and 11 and 13 each have a difference of 2. These pairs are called *twin primes*. Complete the list of all twin primes less than 100. Also, find the sums and products of these twin primes.

Twin Primes	Sums	Products
3 and 5	_____	_____
5 and 7	_____	_____
11 and 13	_____	_____
___ and ___	_____	_____
___ and ___	_____	_____
___ and ___	_____	_____
___ and ___	_____	_____
___ and ___	_____	_____

1. Do you see a pattern in the column of sums? Can you *prove* a fact about the *sums* of twin primes?

2. Do you see a pattern in the column of products? Can you *prove* a fact about the *products* of twin primes?

3. Examine the primes larger than 5 in your list. What *different* digits appear in the units position of these primes? _____ Will any prime larger than 100 have a different ending than the ones you have found? _____

The number 13 is a prime, and 31, the reverse of 13, is also a prime. 13 is called an *emirp* (*prime* spelled backward) because its reverse is a *different prime*. 31 is also an emirp. But 11 is not an emirp. Why not? _____

4. List all the emirps less than 100. _____

Prime Concerns

Over the centuries we have learned a great deal about prime numbers. But in all this time no one has discovered a simple formula that will produce all the primes starting with 2. Many attempts have been made, and no doubt will continue to be made, to find such a formula. One such attempt produced the following:

$$p_n = n^2 - n + 41$$

p_n is a prime number for $n = 1$ through $n = 40$. For example, $p_1 = 41$, $p_2 = 43$, and $p_3 = 47$. Use your calculator or write a computer program to find additional values of p_n for $n = 4, \ldots, 40$. What is p_{41}? _____ Is it prime or composite? Why? _____

5. What are some other values of $n > 41$ for which p_n is a composite number? _____

6. Is p_n prime for some values of $n > 41$? If so, list some of them. _____

Teaching with *Student Math Notes*: Volume 2

Can You . . .

- replace each blank with a "+" or "−" to get an equality relation?

    17 = 1 ___ 2 ___ 3 ___ 5 ___ 7 ___ 11 ___ 13 ___ 13

    19 = 1 ___ 2 ___ 3 ___ 5 ___ 7 ___ 11 ___ 13 ___ 17

    23 = 1 ___ 2 ___ 3 ___ 5 ___ 7 ___ 11 ___ 13 ___ 17 ___ 19 ___ 19

    29 = 1 ___ 2 ___ 3 ___ 5 ___ 7 ___ 11 ___ 13 ___ 17 ___ 19 ___ 23

- find some emirps that contain three, four, or more digits?
- in Eratosthenes' sieve, name the primes whose multiples *must* be crossed out to find all primes less than 200? 300? 500?
- write a computer program based on Eratosthenes' sieve to find all primes less than 1000?
- find a string of at least 1 million consecutive numbers that are all composite? For example, 24, 25, 26, 27, 28 is a string of five consecutive numbers that are all composite. Identify your string by naming the first number and the one-millionth number in the string. (*Hint:* Use factorials!)

Did You Know That . . .

- a conjecture states that the number of twin primes is infinite? No one has been able to prove or disprove this conjecture.
- J. P. Kulik spent twenty years, unassisted, computing a factor table of the numbers from 1 to 100 000 000? He completed his monumental work in 1867. It filled eight volumes, but volume 2 is now missing from the collection at the Vienna Royal Academy.
- a special kind of prime number, *Mersenne numbers,* are named after the French priest and amateur mathematician Marin Mersenne? In 1644, he published his (incomplete) list of primes that satisfied the rule $M_p = 2^p - 1$, where p is prime: p = 2, 3, 5, 7, 13, 17, 19, 31, 67, 127, 257. It took 304 years to resolve the errors in his short list.
- in 1978, the twenty-fifth Mersenne prime, $2^{21701} - 1$, was found by two eighteen-year-old students, Laura Nickel and Curt Noll, using a CYBER-174 computer at California State University at Hayward?
- the largest known Mersenne prime, $2^{216091} - 1$, consists of 65 050 digits and was discovered in 1985 in Houston, Texas, on a Cray X-MP supercomputer by scientists at Chevron Geosciences Company? This find will probably be declared the thirtieth Mersenne prime.
- Christian Goldbach, 1690–1764, conjectured that every even number greater than 4 is equal to the sum of two prime numbers? His conjecture remains unproved today.

NCTM STUDENT MATH NOTES is published as part of the **NEWS BULLETIN** by the National Council of Teachers of Mathematics, 1906 Association Drive, Reston, VA 22091. The five issues a year appear in September, November, January, March, and May. Pages may be reproduced for classroom use without permission.

Editor: Lee E. Yunker, West Chicago Community High School, West Chicago, IL 60185
Editorial Panel: Daniel T. Dolan, Office of Public Instruction, Helena, MT 59620
Elizabeth K. Stage, Lawrence Hall of Science, University of California, Berkeley, CA 94720
John G. Van Beynen, Northern Michigan University, Marquette, MI 49855
Editorial Coordinator: Joan Armistead
Production Assistants: Ann M. Butterfield, Lynn Westenberg

Printed in U.S.A.

Teacher Notes

Page 7. The prime-number pyramid offers an interesting problem-solving activity. Be certain students understand the conditions and requirements. It may be helpful for some to begin with an array where the numbers are listed in order in each row and then rearranged as needed so that successive pairs have prime sums. Note from the example shown that the prime sums can be the same for different pairs in any given row.

Page 8. Encourage students to do some reading and research on the lives of the Greek mathematicians Euclid and Eratosthenes. Extending the sieve to 1 through 200 would require continuing the process through only the next two primes, 11 and 13.

Page 9. Many interesting properties of prime numbers have been found. The formula $P_n = n^2 - n + 41$ was first introduced by Euler in 1772. Several others are given following the answers below.

Page 10. A Mersenne prime is a prime number in the form $2^p - 1$, where p is prime. In 1992, another Mersenne prime was discovered. This new prime, found using a Cray-2 supercomputer, is $2^{756839} - 1$. The magnitude of this prime number is well reflected by the fact that it contains 227 839 digits. The previous record-holding Mersenne prime, found in 1985, was $2^{216091} - 1$ and consisted of a mere 65 050 digits.

Answers

Page 7.
```
                        1   4   3   2   5   6   7
                    1   2   3   4   7   6   5   8
                1   2   3   4   7   6   5   8   9
            1   2   3   4   7   6   5   8   9  10
        1   4   3   2   5   6   7  10   9   8  11
    1   4   3   2   5   6   7  10   9   8  11  12
1   4   3   2   5   6   7  12  11   8   9  10  13
1   2   5   8   3   4   7  12  11   6  13  10   9  14
1   2   5   8   3   4   7  12  11   6  13  10   9  14  15
```

Page 8. 1. No   2. You do not need to cross out the multiples of 11, since the first one not already crossed out is $11 \times 11 = 121$, which is greater than 100.

2, 3, 5, 7, 11, 13, 17, 19, 23, 29, 31, 37, 41, 43, 47, 53, 59, 61, 67, 71, 73, 79, 83, 89, and 97

Page 9. Twin primes: 3 and 5, 5 and 7, 11 and 13, 17 and 19, 29 and 31, 41 and 43, 59 and 61, 71 and 73
1. Except for $3 + 5$, all are divisible by 12: $(6m - 1) + (6m + 1) = 12m$.
2. Each product is 1 less than the square of the average of the twin primes.
3. 1, 3, 7, 9; No
4. 13, 17, 31, 37, 71, 73, 79, and 97
5. $n = 42, 45, 50, 57, 66, 77, 82, 83, 85, 88, 90, 92, 97$
Only thirteen of the first hundred values of n in the formula do not produce primes.

Page 10.
- $+ - - + - + +$
- $+ - + + + + -$
- $- + + - - + - + +$
- $+ - - + - + - + +$

- Emirps: 337, 733, 1471, 1741, ...
- 2, 3, 5, 7, 11, 13; 2, 3, 5, 7, 11, 13, 17; 2, 3, 5, 7, 11, 13, 17, 19
- Here is a string of 1 million consecutive numbers that are all composite:
  $(10^6 + 1)! + 2, (10^6 + 1)! + 3, (10^6 + 1)! + 4, ..., (10^6 + 1)! + (10^6 + 1)$

For some other explorations with primes, try these:
- List numbers of the form $6a \pm 1$, where $a = 1, 2, 3, 4, 5, ...$

a	1	2	3	4	5	6
$6a \pm 1$	5 7	11 13	17 19	23 25	29 31	35 37

Every prime number greater than 3 will appear in this list, but not every number listed will be prime.

- List every number in the form $(2m - 1)(1 + 2a)$, where $a = 1, 2, 3, 4, 5, ...$ and $m = 2, 3, 4, 5, 6, ...$.
Combine the numbers with all even numbers greater than 2 to get all successive nonprime numbers. The omitted odd numbers greater than 1 form, progressively, all the primes.

Nonprimes		4		6		8	9	10		12		14	15	16		18		20	21	22			24	25
Primes	2		3		5					7		11			13		17				19		23		

Page 12 (Extension).   1. 9; 36%   2. 30.0%, 25.0%, 23.0%, 20.6%, 19.5%, 19.0%
4. As N increases, the percent of prime numbers decreases, asymptotically, toward 0.
For $N = 1000$, $P = 16.8\%$. For $N = 10\,000$, $P = 12.3\%$. For $N = 100\,000$, $P = 9.6\%$.

Extension—THE FIRST 500 PRIMES

2	101	233	383	547	701	877	1049	1229	1429	1597	1783	1993	2161	2371	2579	2749	2957	3187	3373
3	103	239	389	557	709	881	1051	1231	1433	1601	1787	1997	2179	2377	2591	2753	2963	3191	3389
5	107	241	397	563	719	883	1061	1237	1439	1607	1789	1999	2203	2381	2593	2767	2969	3203	3391
7	109	251	401	569	727	887	1063	1249	1447	1609	1801	2003	2207	2383	2609	2777	2971	3209	3407
11	113	257	409	571	733	907	1069	1259	1451	1613	1811	2011	2213	2389	2617	2789	2999	3217	3413
13	127	263	419	577	739	911	1087	1277	1453	1619	1823	2017	2221	2393	2621	2791	3001	3221	3433
17	131	269	421	587	743	919	1091	1279	1459	1621	1831	2027	2237	2399	2633	2797	3011	3229	3449
19	137	271	431	593	751	929	1093	1283	1471	1627	1847	2029	2239	2411	2647	2801	3019	3251	3457
23	139	277	433	599	757	937	1097	1289	1481	1637	1861	2039	2243	2417	2657	2803	3023	3253	3461
29	149	281	439	601	761	941	1103	1291	1483	1657	1867	2053	2251	2423	2659	2819	3037	3257	3463
31	151	283	443	607	769	947	1109	1297	1487	1663	1871	2063	2267	2437	2663	2833	3041	3259	3467
37	157	293	449	613	773	953	1117	1301	1489	1667	1873	2069	2269	2441	2671	2837	3049	3271	3469
41	163	307	457	617	787	967	1123	1303	1493	1669	1877	2081	2273	2447	2677	2843	3061	3299	3491
43	167	311	461	619	797	971	1129	1307	1499	1693	1879	2083	2281	2459	2683	2851	3067	3301	3499
47	173	313	463	631	809	977	1151	1319	1511	1697	1889	2087	2287	2467	2687	2857	3079	3307	3511
53	179	317	467	641	811	983	1153	1321	1523	1699	1901	2089	2293	2473	2689	2861	3083	3313	3517
59	181	331	479	643	821	991	1163	1327	1531	1709	1907	2099	2297	2477	2693	2879	3089	3319	3527
61	191	337	487	647	823	997	1171	1361	1543	1721	1913	2111	2309	2503	2699	2887	3109	3323	3529
67	193	347	491	653	827	1009	1181	1367	1549	1723	1931	2113	2311	2521	2707	2897	3119	3329	3533
71	197	349	499	659	829	1013	1187	1373	1553	1733	1933	2129	2333	2531	2711	2903	3121	3331	3539
73	199	353	503	661	839	1019	1193	1381	1559	1741	1949	2131	2339	2539	2713	2909	3137	3343	3541
79	211	359	509	673	853	1021	1201	1399	1567	1747	1951	2137	2341	2543	2719	2917	3163	3347	3547
83	223	367	521	677	857	1031	1213	1409	1571	1753	1973	2141	2347	2549	2729	2927	3167	3359	3557
89	227	373	523	683	859	1033	1217	1423	1579	1759	1979	2143	2351	2551	2731	2939	3169	3361	3559
97	229	379	541	691	863	1039	1223	1427	1583	1777	1987	2153	2357	2557	2741	2953	3181	3371	3571

1. How many numbers from 1 through 25 are prime? Express the answer as a percent.

2. Use the table of primes. Find the percent of the numbers in each interval that are prime.

Interval	1–50	1–100	1–150	1–200	1–250	1–300	1–350	1–400	1–450	1–500
Percent prime	___	___	___	___	___	___	___	___	___	___

3. Draw a graph of the data collected in question 2.

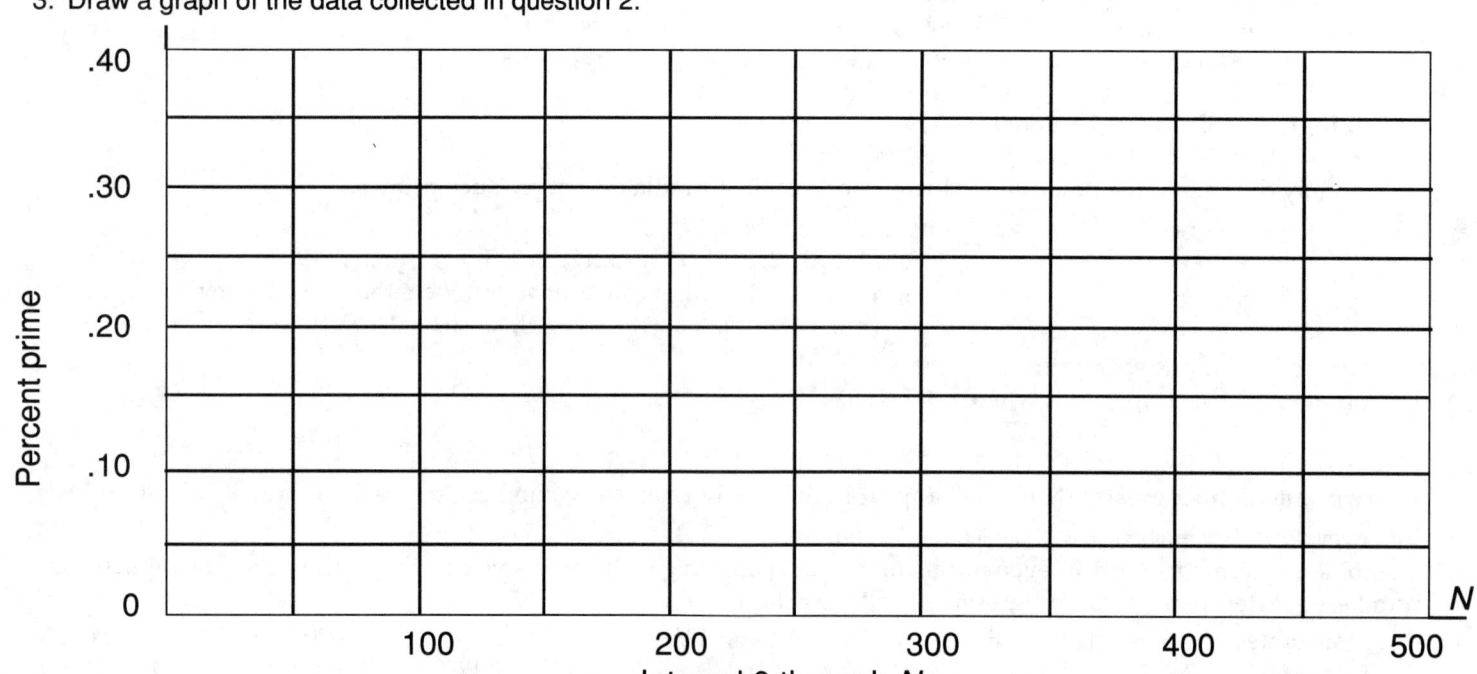

4. As the number N increases, what appears to happen to the percent of prime numbers in the interval from 1 to N?

Models: The Mathematician's Laboratory Equipment

JANUARY 1987

A real-world happening or an abstract idea can be represented by diagrams, sentences, numerals, equations, and graphs. These are called *models*. Can you match the statement on the right with the appropriate model on the left?

a.

_____ 1. Water temperature when the hot-water faucet is turned on and left on

b.

_____ 2. I am thinking of a number. Twice the number minus the number plus 10 minus the number is what?

c. $\underline{V}$, 5, 𝍦

d.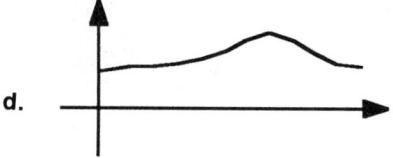

_____ 3. A method of determining the games and winning team in a single-elimination tournament

e. $2n - n + 10 - n = ?$

_____ 4. Fiveness

f.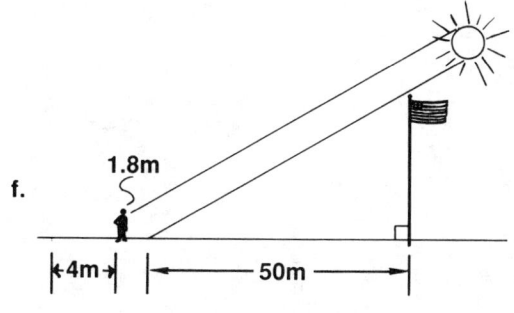

_____ 5. Model of the ratio 4 to 7

_____ 6. An indirect method of measuring height

g.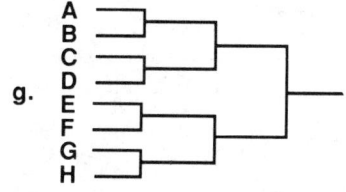

_____ 7. 40 percent

The editors wish to thank Glenn Allinger and Lyle Andersen, Montana State University, Bozeman, MT 59717, for writing this issue of the *NCTM Student Math Notes*.

You use models to—

- find 13 × 24:

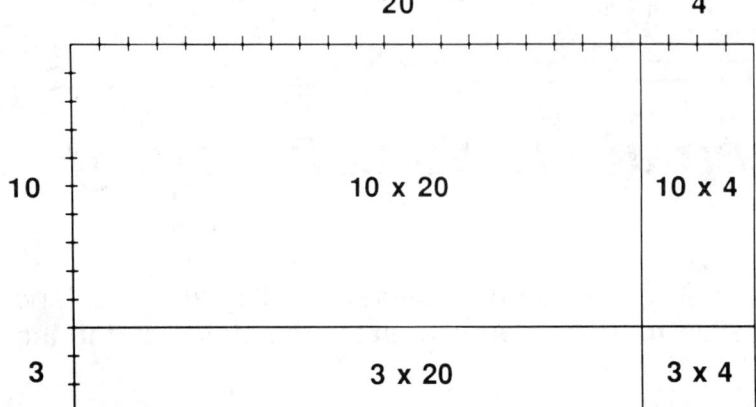

- calculate the product of 2/5 and 1/3:

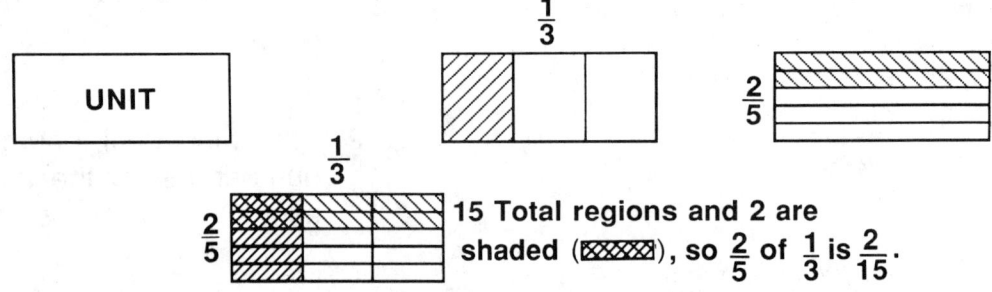

15 Total regions and 2 are shaded (▨), so $\frac{2}{5}$ of $\frac{1}{3}$ is $\frac{2}{15}$.

- determine 25 percent of 60:

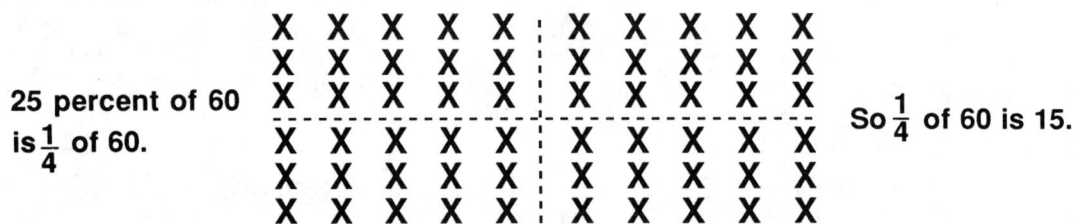

25 percent of 60 is $\frac{1}{4}$ of 60.   So $\frac{1}{4}$ of 60 is 15.

Related problems

Sketch a model and use it to calculate the following products:

8. Total number of people in 34 rows of 29 people each

9. 3/4 of 5/7

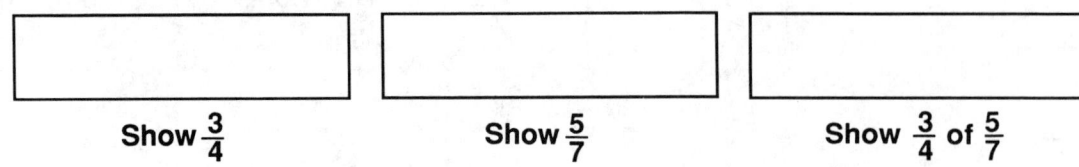

Show $\frac{3}{4}$ Show $\frac{5}{7}$ Show $\frac{3}{4}$ of $\frac{5}{7}$

10. A ski shop advertises 10 percent off on a pair of thermal gloves originally priced at $40. What is the dollar discount?

11. $(x + 3)(x + 4)$

Teaching with *Student Math Notes:* Volume 2

Many Problems Have the Same Model

Sometimes one problem seems very different from another problem, yet the models used to help determine the solution are basically the same.

Example 1: Rosita enters a room in which Chad, Dawn, and Paul are already present. Everyone in the room shakes hands with everyone else. How many total handshakes will take place?

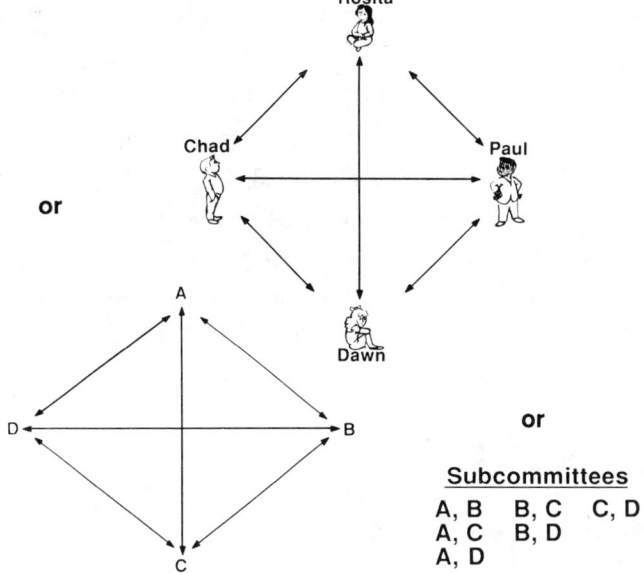

Rosita, Paul
Rosita, Dawn   Paul, Dawn
Rosita, Chad   Paul, Chad   Dawn, Chad   or

Example 2: Four sixth graders decide to help clean up the school grounds. They form subcommittees of two people each, with each subcommittee responsible for recycling a certain material, for example, metal, paper, glass, and chemicals. How many different subcommittees are possible?

or

Subcommittees
A, B   B, C   C, D
A, C   B, D
A, D

Related problems

12. Five Girl Scouts from different troops meet for a leadership conference. In introducing themselves to each other, every girl shakes the hand of each of the other scouts. How many total handshakes are made?

13. From a group of five aerobic exercises, you can choose to do any three during the next free exercise period. The exercises require the stretching of muscles in the ankle, thigh, calf, shoulders, or neck. How many different groups of three exercises are possible?

14. Kirstin has placed five points on a circle. How many different triangles can she draw—counting *only* those triangles that have three of the original points as vertices?

15. Can you solve this problem by graphing?

> A hiker climbed a mountain one day, starting at 6:00 a.m. and arriving at the 3740-meter summit at about 3:00 p.m. She took pictures all day and camped out that night. She began the return journey at 6:00 the next morning and arrived at the base at noon. Was there some point on the trail that she reached at the *same* time when both climbing and descending?

Teaching with *Student Math Notes:* Volume 2

Did You Know That...

- in the time of Columbus, many people thought that a flat plane was a good model for representing the earth?
- every time you solve a mathematical problem with pencil and paper you are building a model?
- the game of chess is a model of ancient war games?
- physicists think that the geometric model best suited to represent the universe is not Euclidean but elliptic, in which there are no parallels through a point outside a line?
- computers can produce photographic models using digits? The following digits represent a portion of a weather-satellite photograph transmitted by short-wave radio: 18462525747838383 . . .
- Japanese fishing crews use computer simulations to determine where the best fishing will be?
- meteorologists use the equation $D^3 = 216T^2$ as a model to describe the size and intensity of four types of violent storms: tornadoes, thunderstorms, hurricanes, and cyclones? D is the diameter of the storm in miles and T is the number of hours the storm travels before dissipating.

Can You...

16. use the equation developed by meteorologists and your calculator to answer the following?

    a. The world's worst recorded monsoon took place on 13–14 November 1970 in the Ganges delta islands in Bangladesh. More than 1 000 000 people died. If this storm lasted 24 hours, what was its diameter?

    b. In the U.S., about 150 tornadoes occur each year, mostly in the central plains states, with the highest frequency in Iowa, Kansas, Arkansas, Oklahoma, and Mississippi. If a tornado's diameter is 2 miles, how long would it be expected to last?

17. make a model to represent the product $(2x + 1)(3x + 2)$? (*Hint:*  represents x^2.)

18. create a model to predict the minimum number of moves necessary for $K/2$ boys sitting in the front half of a canoe to exchange seats with $K/2$ girls sitting in the back half of a canoe with one empty seat between them given the following conditions?

    (a) K is an even natural number.
    (b) Each seat can hold only one person.
    (c) The only legal moves are to an adjacent empty seat or to step over one person to an empty seat.
    (d) The canoe will always have one more seat than people.

When the moves are finished, the boys will have moved to the back half of the canoe and the girls to the front half. The seat in the middle will be empty.

Teacher Notes

Page 13. After students match the seven models with their appropriate statements, consider modifying the statements and having students draw and discuss new models. These new models might include the following:
- Water temperature when cold water faucet is turned on and left on
- Twice a number plus that number minus 10 plus that number is what?
- Model of the ratio 4/7
- A representation of 400 percent

Page 14. The models described here are pictorial. Hands-on versions made from folding or cutting paper make the processes more concrete and visual.

Page 15. Many problems can have the same model as that shown in the two examples. Similarly, many problems can have more than one appropriate model.

Answers

Page 13. 1. *d* 2. *e* 3. *g* 4. *c* 5. *a* 6. *f* 7. *b*

Page 14. 8.

	30	4
20	20 × 30	20 × 4
9	9 × 30	9 × 4

$$\begin{array}{r} 29 \\ \times 34 \\ \hline 36 \\ 80 \\ 270 \\ 600 \\ \hline 986 \end{array}$$

9. $\frac{5}{7}$

$\frac{3}{4}$

$\frac{3}{4} \times \frac{5}{7} = \frac{15}{28}$

10. [grid of $ signs]

11.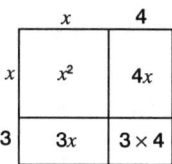

$x^2 + 3x + 4x + 12$
$= x^2 + 7x + 12$

Page 15. 12. 10 handshakes   13. 10 different groups   14. 10 triangles

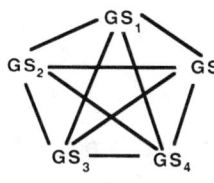  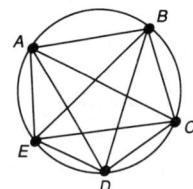

15.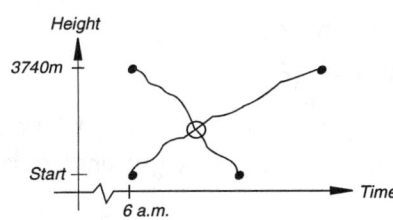

Page 16. 16. a) Approximately 50 miles $d^3 = 216(24)^2$ $d = 49.922$
b) Approximately 0.19 hours $2^3 = 216t^2$ $t = 0.19245$

17.

	x	*x*	*x*	1	1
x	x^2	x^2	x^2	*x*	*x*
x	x^2	x^2	x^2	*x*	*x*
1	*x*	*x*	*x*	1	1

$6x^2 + 7x + 2$

18. (Extension)

When $K = 2$, the solution will have three moves (G stands for girls, B for boys): G___B, GB___, ___BG, B___G. When $K = 4$, the solution will have eight moves: GG___BB, GGB___B, G___BGB, ___GBGB, BG___GB, BGBG___, BGB___G, B___BGG, BB___GG. When $K = 6$, the solution will have fifteen moves. When $K = 8$, the solution will have twenty-four moves. In general, for any K, the minimum number of moves will be $(1/4)K^2 + K$.

Answers, continued

Page 18.
1. Eight; two; $a \times a \times a, b \times b \times b$
2. Square rectangular prisms; three at $a \times a \times b$ and three at $a \times b \times b$
3. $(a + b)^3 = a^3 + 3a^2b + 3ab^2 + b^3$
4. $P(LSY) = 6.4\%$ $P(LSR) = 9.6\%$ $P(LDY) = 14.4\%$ $P(LDR) = 14.4\%$
   $P(TSY) = 9.6\%$ $P(TSY) = 9.6\%$ $P(TDY) = 14.4\%$ $P(TDR) = 21.6\%$

   Note that the sum of all eight possibilities is 100 percent.
   $$(a + b)^3 = a^3 + 3a^2b + 3ab^2 + b^3 = 6.4\% + 3(9.6\%) + 3(14.4\%) + 21.6\% = 100\%$$

Extension—MODELING IN THREE DIMENSIONS

A cube is cut through in three different directions as shown.

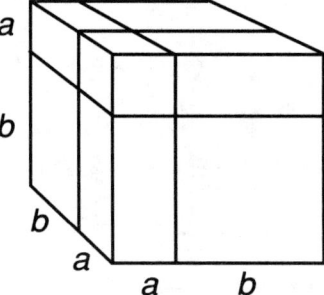

1. How many pieces will be formed?
   How many of the pieces will be cubes?
   Give the dimensions of each new cube formed.

2. Name the shapes of the remaining pieces.
   Give the dimensions of each one.

3. The pieces formed by cutting a cube this way produce four sets of different sizes and shapes. Combined this way, the four corresponding volumes form a concrete model for $(a + b)^3$. Complete the algebraic equivalence.

   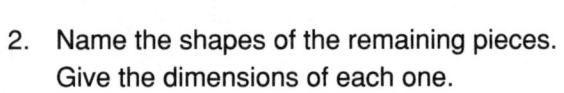

Now use this model to solve a new problem.

4. An assortment of seeds for a certain flowering plant has the following three independent characteristics:

   - 40% will produce one large flower (L). The rest will have many tiny flowers (T).
   - 40% will produce single-petaled flowers (S). The rest will be double petaled (D).
   - 40% will produce yellow flowers (Y). The rest will have red flowers (R).

   Let all seeds in the assortment be represented by the entire cube. Let the cube measure 1.00 on each edge with $a = 0.40$ and $b = 0.60$. Use the volumes of the eight pieces to find the probabilities for the eight kinds of flowers possible.

Polygonal Numbers

MARCH 1987

Mathematicians have given names to sets of numbers. You may know some of them: whole numbers, integers, natural numbers, real numbers, irrational numbers, and so on. Some numbers are associated with polygons and have geometric names.

The dots in the following figures represent *geometric numbers.* Fill in the blanks and determine each number.

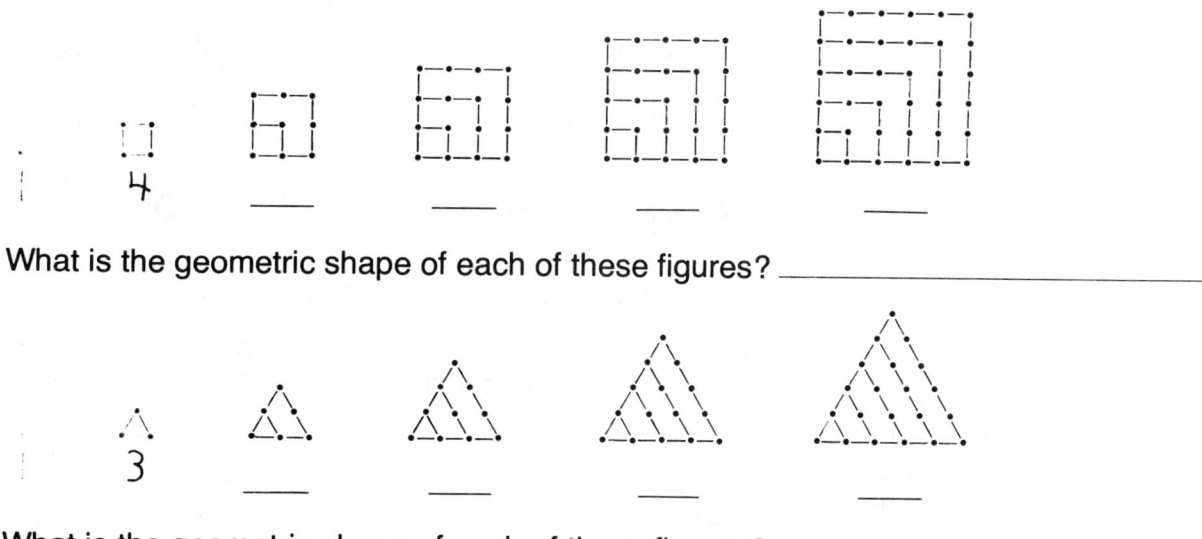

1   4   ___   ___   ___

What is the geometric shape of each of these figures? _____

___   3   ___   ___   ___   ___

What is the geometric shape of each of these figures? _____

The numbers determined by the shapes in the first group are called *square numbers.* The numbers determined by the shapes in the second group are called *triangular numbers.*

Mathematicians use abbreviations or symbols to represent geometric numbers. For example, we write S_4 to represent the fourth square number, a square array with four dots on each side. Similarly we can write T_3 to represent the third triangular number, a trianglar array with three dots on each side. Using this notation, we can write $S_5 = 25$ and $T_3 = 6$.

Several interesting patterns can be explored involving triangular and square numbers, so you will need to build a table to show what some of these numbers are.

n	1	2	3	4	5	6	7	8	9	10	11	12
S_n	1	4	9	16	25	36						
T_n	1	3	6	10	15	21						

n	13	14	15	16	17	18	19	20	21	22	23	24
S_n												576
T_n												300

The editors wish to thank Boyd Henry, College of Idaho, Caldwell, ID 83605, for writing this issue of the *NCTM Student Math Notes.*

Explorations with Triangular Numbers

- Find the sum of any two consecutive triangular numbers. For example, suppose we pick T_6 and T_7; $21 + 28 = 49$. Their sum is 49, the seventh square number. Add T_8 and T_9. Is the sum of these two consecutive triangular numbers a square number? What do you think you can say about the sum of *any* two consecutive triangular numbers?

- Draw an array to represent the seventh square number, S_7.
  Can you separate it into two consecutive triangular numbers?

- Complete the table below.

	n	1	2	3	4	5	6	7	8	9	10
A	T_n	1	3	6	10						
B	$8 \times T_n$	8	24	48							
C	$[8 \times T_n] + 1$	9	25	49							

What kind of numbers are in row C? In general, if we select any triangular number, multiply it by 8, and add 1, what is true about the result? _____

Further Explorations

- Let's investigate the nature of fractions that have numerators and denominators consisting of consecutive square numbers and consecutive triangular numbers. But first, complete the following table.

n	1	2	3	4	5	6	7	8	9
S_n/S_{n+1}	1/4	4/9	9/16						
T_n/T_{n+1}	1/3	3/6	6/10						

Look carefully at the fractions in the first row. All the numerators and denominators are square numbers. Can any of these fractions be reduced to lower terms?

Now examine the fractions in the second row, in which the numerators and denominators are all triangular numbers. The first fraction, 1/3, cannot be reduced. But what about the other fractions? Can all of them be reduced to lower terms?

Continuing Our Explorations

- As you complete the following table, consider these questions:

  What happens when a number is added to its square and the sum is divided by 2?
  (This is simply the average of any number and its square.)

  What happens when a number is subtracted from its square and the difference is divided by 2?

	n	2	3	4	5	6	7	8	9	10	11	12
	S_n	4	9	16								
	$S_n + n$	6	12	20								
A	$(S_n + n)/2$	3	6	10								
	$S_n - n$	2	6	12								
B	$(S_n - n)/2$	1	3	6								

- What do you notice about the numbers in rows A and B? _____

Explorations with Cubical Arrays

- We have already observed that a square number can be represented with a square array. In a similar manner, we can represent numbers called *cubes* with a cubical array. To build a cubical array measuring four units on each side, we need sixty-four units, since each of the four layers of the cube is a square array measuring 4×4. If we use the notation C_n to mean the cube of n, then $C_n = n^3$. Completing this table will help you recognize numbers that are cubes.

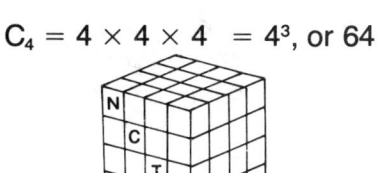

$C_4 = 4 \times 4 \times 4 = 4^3$, or 64

n	1	2	3	4	5	6	7	8	9	10	11	12
C_n	1	8	27	64								

- Now we will explore both the sum and difference of the squares of two consecutive triangular numbers. But first you must complete the following table.

	n	1	2	3	4	5	6	7	8	9	10
	T_n	1	3	6							
	T_{n+1}	3	6	10							
A	$[T_n]^2$	1	9	36							
B	$[T_{n+1}]^2$	9	36	100							
C	Row B + Row A	10	45	136							
D	Row B − Row A	8	27	64							
E	S_{n+1}	4	9	16							
F	$T_{S_{n+1}}$	10	45	136							

- In row C we have the sums of the squares of consecutive triangular numbers. Is each entry in row C a triangular number? _____ We know that the entries in row F are triangular numbers; in fact the entries in row F are triangular numbers associated with square numbers. Compare row C with row F. What do you conclude about the sum of the squares of two consecutive triangular numbers? _____

- In row D we have the difference between the squares of two consecutive triangular numbers. What kind of numbers do we have in row D? _____

- Let's explore some other relationships by completing this table.

	n	1	2	3	4	5	6	7	8	9	10
A	C_{S_n}	1	64	729							
B	S_{C_n}	1	64	729							
C	T_{S_n}	1	10	45							
D	S_{T_n}	1	9	36							
E	Row A − Row B	0	0	0							
F	Row C − Row D	0	1	9							

- What conclusions can you draw from row E? _____
- What conclusions can you draw from row F? _____

Teaching with *Student Math Notes:* Volume 2

Did You Know That...

- There are pentagonal and hexagonal numbers as well as triangular and square numbers? In fact numbers exist for each polygon. They are called *polygonal* or *figurate* numbers. Many interesting patterns can be found among these numbers.

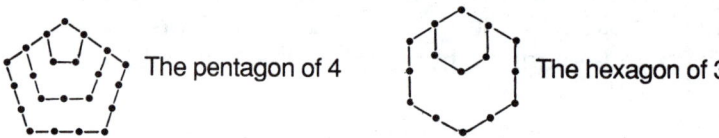

The pentagon of 4   The hexagon of 3

- This BASIC program will compute any polygonal or figurate number you desire.

```
10   PRINT "THIS PROGRAM LISTS THE N-GON OF EACH NUMBER FROM 1 TO K"
20   INPUT " SELECT N ( ANY WHOLE NUMBER 3 OR GREATER).";N
30   INPUT "HOW FAR DO YOU WANT THE LIST TO GO";K
40   FOR A = 1 TO K
50   PRINT A;TAB(15);((N−2)∗A∗A−(N−4)∗A)/2
60   NEXT A
70   END
```

Can You . . .

- Find patterns among the polygonal numbers by completing the following table?

n	1	2	3	4	5	6	7	8	9	10
Triangle: $T_n = n(n+1)/2$	1	3	6							
Square: $S_n = n^2$	1	4	9							
Pentagon: $P_n = n(3n-1)/2$	1	5	12							
Hexagon: $H_n = n(4n-2)/2$	1	6	15							
Heptagon: $HP_n = n(5n-3)/2$	1	7	18							
Octagon: $O_n = n(6n-4)/2$	1	8	21							

- Show that . . .

S_n/S_{n+1} for $n \geq 1$ cannot be reduced to lower terms?

T_n/T_{n+1} for $n > 1$ can always be reduced to lower terms?

- Prove by mathematical induction that . . .

$T_n + T_{n+1} = S_{n+1}$ for $n \geq 1$?

$[8T_n + 1] = (2n+1)^2$ for $n \geq 1$?

$(S_n + n)/2 = T_{n+1}$ for $n \geq 2$?

$(S_n - n)/2 = T_n$ for $n \geq 2$?

$[T_n]^2 + [T_{n+1}]^2 = T_{S_{n+1}}$ for $n \geq 1$?

$[T_{n+1}]^2 - [T_n]^2 = C_{n+1}$ for $n \geq 1$?

NCTM STUDENT MATH NOTES is published as part of the **NEWS BULLETIN** by the National Council of Teachers of Mathematics, 1906 Association Drive, Reston, VA 22091. The five issues a year appear in September, November, January, March, and May. Pages may be reproduced for classroom use without permission.

Editor: Lee E. Yunker, West Chicago Community High School, West Chicago, IL 60185
Editorial Panel: Daniel T. Dolan, Office of Public Instruction, Helena, MT 59620
Elizabeth K. Stage, Lawrence Hall of Science, University of California, Berkeley, CA 94720
John G. Van Beynen, Northern Michigan University, Marquette, MI 49855
Editorial Coordinator: Joan Armistead
Production Assistants: Ann M. Butterfield, Karen Aiken

Printed in U.S.A.

Teacher Notes

Page 19. The activity sheet on page 24 can be used in conjunction with page 19. The activity sheet has students construct successive polygonal numbers of various types on square dot paper. Watch for students who forget to include or count the inside dots in the various arrays.

Page 20. Here students generate and explore number patterns to strengthen their problem-solving skills. Encourage them to discover and express other number relationships that they see as well.

Page 21. Some students may need help in understanding the subscripts used in the table at the bottom of the page. C_{S_n} is the cube of the square of n, and S_{C_n} is the square of the cube of n.

Page 22. Students may see the number relationships for pentagonal, hexagonal, heptagonal, and octagonal numbers better through a geometric, rather than a numerical, vantage point. In this case, you may choose to use the activity sheet on page 24 first.

Answers

Page 19.   Square numbers: $S_n = 1, 4, 9, 16, 25, 36, \ldots$
              Triangular numbers: $T_n = 1, 3, 6, 10, 15, 21, \ldots$

n	7	8	9	10	11	12	13	14	15	16	17	18	19	20	21	22	23
S_n	49	64	81	100	121	144	169	196	225	256	289	324	361	400	441	484	529
T_n	28	36	45	55	66	78	91	105	120	136	153	171	190	210	231	253	276

Page 20.
- The sum of two successive triangular numbers, $T_n + T_{n+1}$, is the square number S_{n+1}.
- $S_7 = 49 = 21 + 28 = T_6 + T_7$
-

4	5	6	7	8	9	10
10	15	21	28	36	45	55
80	120	168	224	288	360	440
81	121	169	225	289	361	441

For any triangular number T_n, $8T_n + 1$ is the square number S_{2n+1}.

-

4	5	6	7	8	9
16/25	25/36	36/49	49/64	64/81	81/100
10/15	15/21	21/28	28/36	36/45	45/55

The ratios of successive square numbers can never be reduced.
The ratios of successive triangular numbers can always be reduced.

-

5	6	7	8	9	10	11	12
25	36	49	64	81	100	121	144
30	42	56	72	90	110	132	156
15	21	28	36	45	55	66	78
20	30	42	56	72	90	110	132
10	15	21	28	36	45	55	66

$(S_n + n)/2$ and $(S_n - n)/2$ always yield triangular numbers T_n and T_{n-1}.

Teaching with *Student Math Notes*: Volume 2

Page 21.

	5	6	7	8	9	10	11	12
	25	36	49	64	81	100	121	144

	4	5	6	7	8	9	10
	10	15	21	28	36	45	55
	15	21	28	36	45	55	66
	100	225	441	784	1296	2025	3025
	225	441	784	1296	2025	3025	4356
	325	666	1225	2080	3321	5050	7381
	125	216	343	512	729	1000	1331
	25	36	49	64	81	100	121
	325	666	1225	2080	3321	5050	7381

- $[T_{n+1}]^2 + [T_n]^2 = T_{S_{n+1}}$
- $[T_{n+1}]^2 - [T_n]^2 = (n+1)^3$

	4	5	6	7	8	9	10
	4 096	15 625	46 656	117 649	262 144	531 441	1 000 000
	4 096	15 625	46 656	117 649	262 144	531 441	1 000 000
	136	325	666	1 225	2 080	3 321	5 050
	100	225	441	784	1 296	2 025	3 025
	0	0	0	0	0	0	0
	36	100	225	441	784	1 296	2 025

- $C_{S_n} - S_{C_n} = 0$
- $T_{S_n} - S_{T_n} = [T_{n-1}]^2$

Page 22.

	4	5	6	7	8	9	10
	10	15	21	28	36	45	55
	16	25	36	49	64	81	100
	22	35	51	70	92	117	145
	28	45	66	91	120	153	190
	34	55	81	112	148	189	235
	40	65	96	133	176	225	280

Successive differences in column n are constant and equal to the triangular number T_n.

Page 25 (Extension).

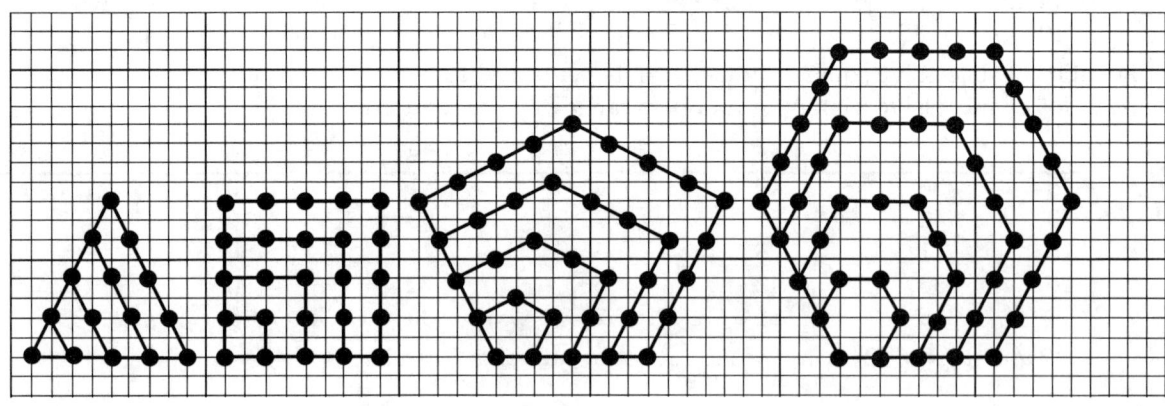

Page 26. Decagonal numbers: 1, 10, 27, 52, ...

Extension—DRAWING FIGURATE NUMBERS

Successive figurate numbers of a given class can be constructed on square dot paper by replicating the shape to a scale 1 greater at each new stage. Complete the next two stages of these triangular, square, pentagonal, and hexagonal numbers.

TRIANGULAR NUMBERS

SQUARE NUMBERS

PENTAGONAL NUMBERS

HEXAGONAL NUMBERS

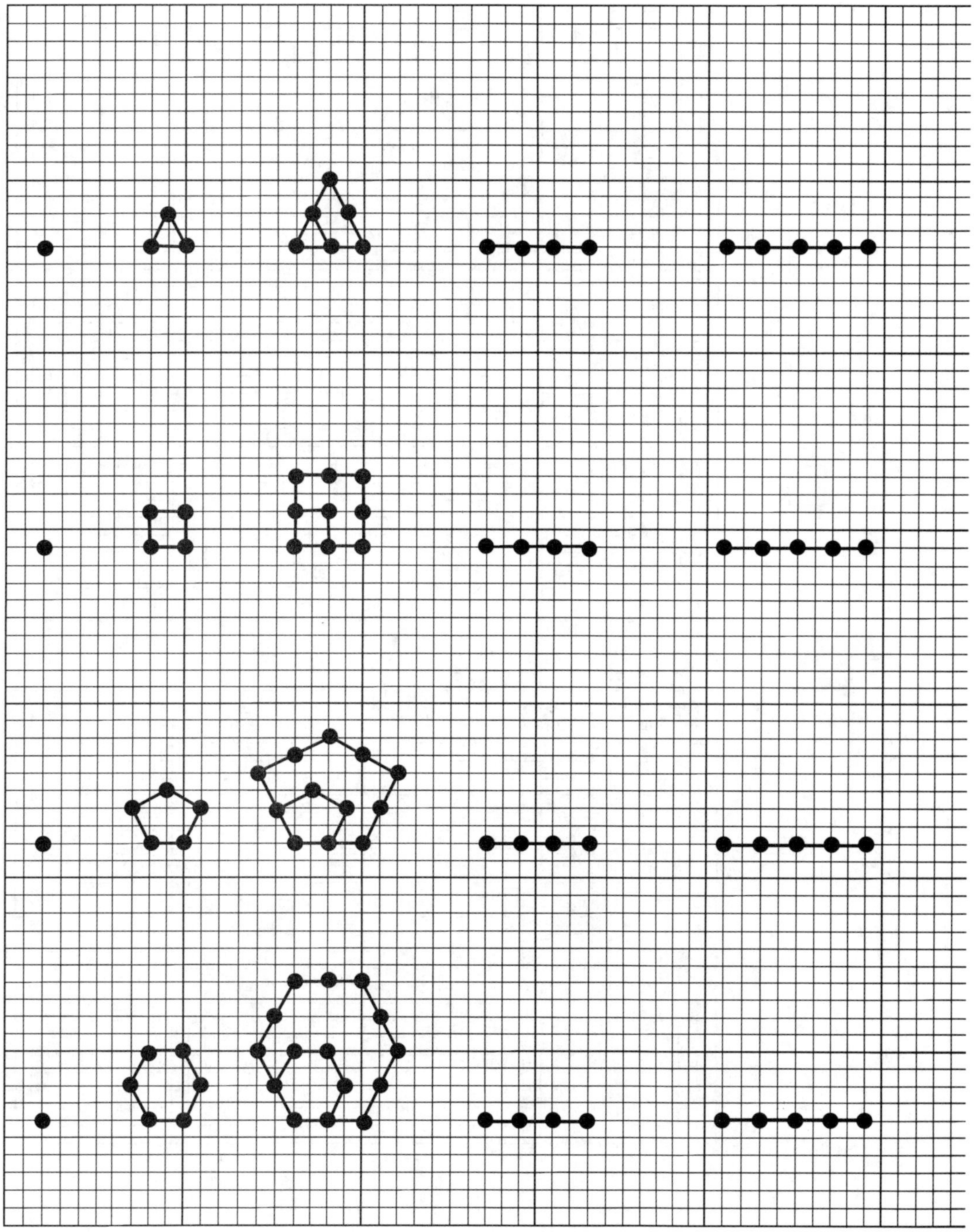

Teaching with *Student Math Notes*: Volume 2

The first two decagonal numbers are 1 and 10. Draw a figure that can be used to represent the next decagonal number. Then count all its dots.

Fair Games

Playing games is a frequent source of entertainment. We enjoy the stimulation, the challenge to find winning strategies, and the competition. The following activities will allow you to explore three different games involving two players.

Game 1

Use a marker and the game board below. The rules for play are as follows:

1. Player A places the marker in any one of the empty cells in the top row.
2. Player B moves the same marker one cell to the right, one cell to the left, or one cell straight down. A player is *not* allowed to move up or diagonally.
3. Players alternate turns.
4. A player is *not* allowed to move to the previously occupied cell.
5. The first player who moves the marker into the winning area wins the game.

Play the game several times with another student.

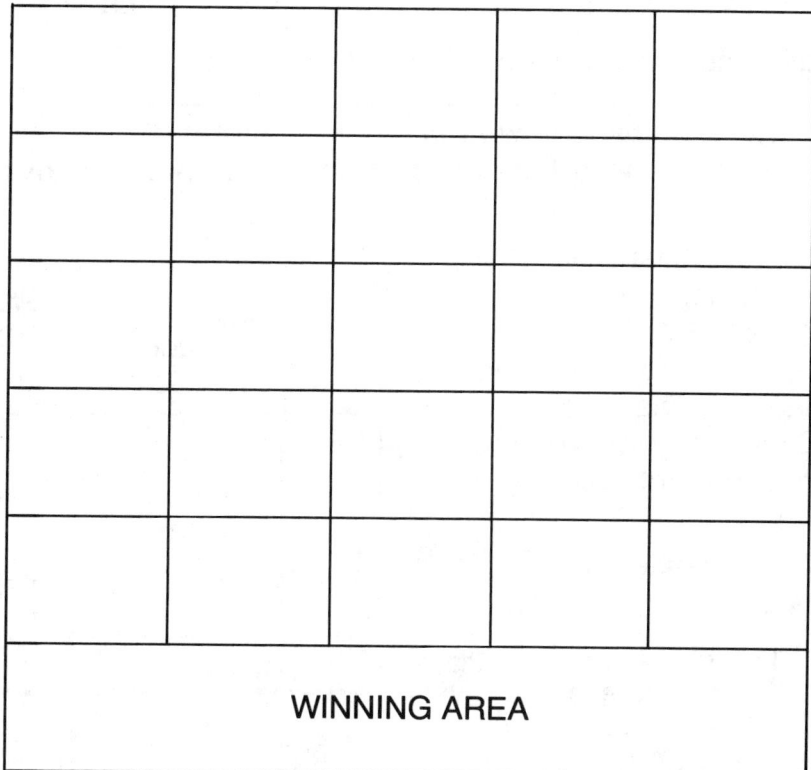

Can you find a winning strategy? Does one appear to gain any advantage by being first or second? Analyze the game together with another student. Try to find a winning strategy. Find a way so that you can win every time.

Analyzing Fairness of Games

> **Definition:** A game is fair if each player has a fifty-fifty chance of winning, or the probability of winning is one-half.

Game 2 (Version 1)

Here is a very basic but fun game for two people.

To play the game one person is called EVEN and the other is called ODD. Each person secretly writes down a number from *0 through 9* and covers the number. Next show the numbers. If the sum of the two numbers is even, the EVEN player wins. If the sum is odd, the ODD player wins. For example, if the two numbers written are 7 and 5, then the sum is 12, so the EVEN player wins.

		Total Wins
EVEN		
ODD		

Play the game several times, keeping track of wins by recording tally marks in the chart.

- Is this a fair game? _____ Argue your position.

Game 2 (Version 2)

A second version of the game is played *without 0*; thus a person can choose only numbers *1 through 9*. Play the game as described above.

		Total Wins
EVEN		
ODD		

- Is this a fair game? _____
- Who has the mathematical advantage, EVEN or ODD? _____

> **Definition:** The probability of winning is the number of favorable outcomes divided by the total number of possible outcomes.

Let's determine the probability that ODD will win. Fill in the addition table. Note that the usual order of headings 1, 2, 3, 4, 5, 6, 7, 8, 9 has been rearranged for convenience. From the table we have 81 possible outcomes. Counting gives 40 with an odd sum. Hence the probability that ODD will win is 40/81.

Let's analyze the probability another way. ODD can win in two ways:

Case 1: EVEN can choose an odd number and ODD can choose an even number. These outcomes are shown in the upper-right-hand corner of the table.

Case 2: EVEN can choose an even number and ODD can choose an odd number. These outcomes are shown in the lower-left-hand corner of the table.

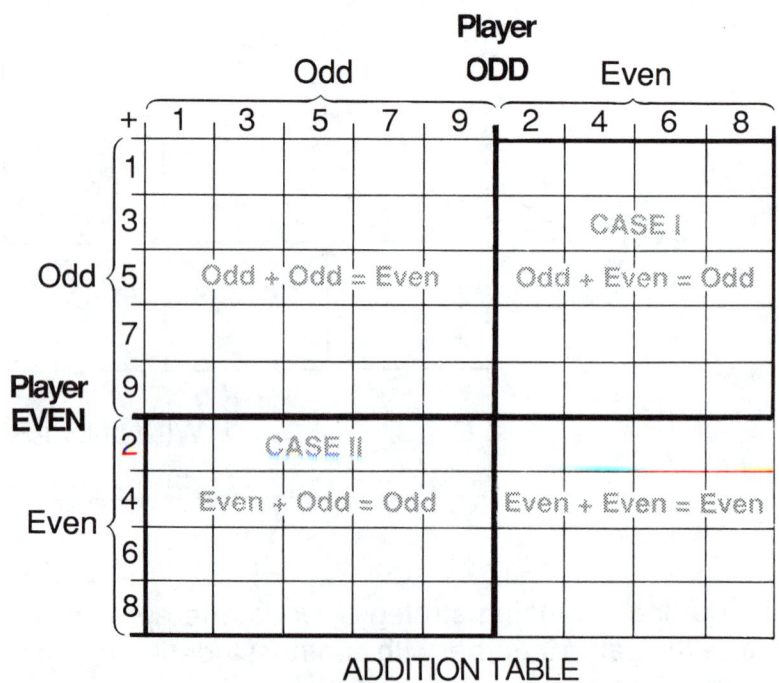

ADDITION TABLE

Let's look further into case 1. The first five rows in the table identify those outcomes where EVEN chose an odd number. This total constitutes 5/9 of 81, or 45, outcomes. The last four columns identify those outcomes where ODD chose an even number. Hence, only 4/9 of EVEN's 45 outcomes satisfy case 1; that is, only 4/9 of 45, which is 20. Therefore, the probability that case 1 will occur is 4/9·5/9, or 20/81.

A parallel argument is used for case 2. The last four rows in the table identify those outcomes where EVEN chose an even number. This amount constitutes 4/9 of 81, or 36, outcomes. The first five columns identify those outcomes where ODD chose an odd number. Hence, only 5/9 of EVEN's 36 outcomes satisfy case 2; that is, 5/9 of 36, which is 20. Therefore, the probability that case 2 will occur is 5/9·4/9, or 20/81.

Since case 1 and case 2 are separate cases, the probability that ODD will win is 4/9 · 5/9 + 5/9 · 4/9 = 20/81 + 20/81 = 40/81.

- What is the probability that EVEN will win? _____

Game 2 (Version 3)

This game can also be played with the same numbers (1–9); however, find the product of the two numbers chosen instead of the sum.

- Is this version a fair game? _____
- If you have a choice would you prefer to be EVEN or ODD? _____

Game 2 (Version 4)

By now you should realize that EVEN has a definite advantage in version 3. In fact, EVEN will always win by writing down any even number, since the product of two even numbers or of an even number and an odd number is always even.

To prevent this eventuality, assume that a number from 1 through 9 is randomly assigned to you and your friend; that is, you no longer get to pick your number.

- What is the probability that the product is odd? _____
- What is the probability that the product is even? _____
- Is the game now "fair"? _____

Game 3

Suppose that two white marbles and two black marbles are in a box. Without looking in the box, randomly choose two of the four marbles. If the two marbles are the same color, player A wins. If the two chosen marbles are each of a different color, player B wins. Play this game several times with another student.

- Is this game fair? _____ Why?

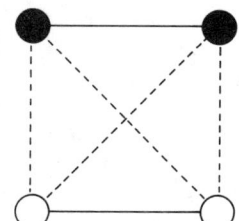

The model should convince you that it is not a fair game. It shows that six different ways exist to draw two of the four marbles. The solid lines show that only two ways exist to draw two marbles of the same color. The dotted lines show that four ways exist to draw two marbles of a different color. "Different" is favored to win by a two-to-one advantage over "same."

This figure shows three black marbles and two white marbles. Draw in solid lines for "same." Draw in dotted lines for "different."

- What is the probability that "same" will win? _____
- What is the probability that "different" will win? _____

• Many combinations of black and white marbles will produce a fair game. Can you find a combination to make it a fair game? _____
• Can you find other combinations that will make it a fair game? _____

Can You...

- write a computer program that will simulate the various versions of game 2 to generate the empirical probabilities?
- write a computer program that will simulate game 3 for various combinations of black and white marbles?

Did You Know That...

- games constitute an ideal model system in the development of artificial intelligence?
- two Frenchmen, Pierre de Fermat and Blaise Pascal, laid the foundations of probability theory through the analysis of various gambling games popular in France in the seventeenth century?
- archeologists uncovered, at prehistoric sites, large numbers of bones called *astragalia* that were apparently used as dice in ancient games?
- in England the game of American checkers is called *draughts*? Polish checkers is played on a ten-by-ten board with twenty pieces per side, rather than the usual twelve pieces. Turkish checkers is played with sixteen pieces per side.
- Johannes Gutenberg, inventor of the printing press, printed playing cards in 1440, the same year he printed his famous Gutenberg Bible?
- The game of chess was first played in Asia and was made popular in the United States by Benjamin Franklin?

NCTM STUDENT MATH NOTES is published as part of the **NEWS BULLETIN** by the National Council of Teachers of Mathematics, 1906 Association Drive, Reston, VA 22091. The five issues a year appear in September, November, January, March, and May. Pages may be reproduced for classroom use without permission.

Editor: Lee E. Yunker, West Chicago Community High School, West Chicago, IL 60185
Editorial Panel: Daniel T. Dolan, Office of Public Instruction, Helena, MT 59620
Elizabeth K. Stage, Lawrence Hall of Science, University of California, Berkeley, CA 94720
John G. Van Beynen, Northern Michigan University, Marquette, MI 49855
Editorial Coordinator: Joan Armistead
Production Assistants: Ann M. Butterfield, Karen K. Aiken

Printed in U.S.A.

Teacher Notes

Page 27. Game 1 is entertaining and challenging. Encourage students to play the game at first just for fun and enjoyment, to become familiar with the rules. They will soon start applying their own problem-solving skills trying to find a winning strategy. For those who need a helpful hint, suggest working backward to plan a win.

Page 28. A fair game must satisfy the given definition by mathematical analysis. Thus, it is an abstract ideal. Empirical results found by actually playing a fair game repeatedly most likely will, in the short term, yield winning results somewhat more or less than half the time.

Page 29. Game 3 conveniently lends itself to both a mathematical analysis through the geometric figure used as a model and an empirical analysis through actual play. The game offers an opportunity for students to collect and analyze experimental data. Students need to see real-life connections to statistics in as many ways as possible in order to develop good data sense as well as number and spatial sense. As teachers, we need to be careful not to overemphasize the abstract mathematical analysis through models. Many practical decisions are made on the basis of actual experiences.

Answers

Page 27. Here is an analysis of game 1. As soon as you place a marker in the bottom row of cells, your opponent will play directly below you and win. Therefore, your strategy is to avoid moving into the last row of cells. Safe cells in the row above it are in the first, third, and fifth positions. Work up in the same way, row by row, to the top row. Through this process, we see that safe cells are the second and fourth in the top row.

Each safe cell is marked with an X in this diagram. If you go first, place your marker in a safe cell in the top row. Play carefully, and you are sure to win.

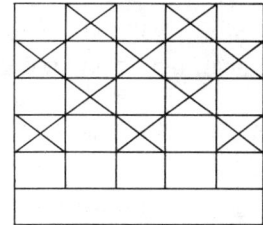

Students may enjoy searching for winning strategies in other arrays with different numbers of rows and columns.

Page 28.   **Game 2** (Version 1):   Not fair. EVEN has a slight advantage.

Page 29.   **Game 2** (Version 2): $P(EVEN) = 1 - P(ODD) = 1 - 40/81 = 41/81$ or
$P(EVEN) = 5/9 \times 5/9 + 4/9 \times 4/9 = 25/81 + 16/81 = 41/81$

Game 2 (Version 3):   Not a fair game. $P(ODD) = 25/81$ and $P(EVEN) = 56/81$

Game 2 (Version 4):   Not a fair game. $P(ODD) = 5/9 \times 5/9 = 25/81 = 0.31$
$P(EVEN) = 4/9 \times 4/9 + 4/9 \times 5/9 + 5/9 \times 4/9 = 56/81 = 0.69$

Game 3: $P(SAME) = 1/3$ and $P(SAME) = 4/10 = 0.4$ and
$P(DIFFERENT) = 2/3$ $P(DIFFERENT) = 6/10 = 0.6$

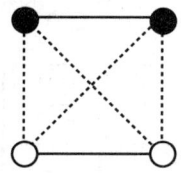 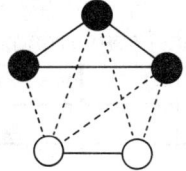

Page 30.   • Use 4 marbles, 3 black and 1 white, for a fair game.
• $P(SAME) = 3/6 = 0.5$ • $P(DIFFERENT) = 3/6 = 0.5$
Two other choices are 16 marbles—6 black and 10 white, or 10 black and 6 white.
$P(SAME) = 6/16 \times 5/15 + 10/16 \times 9/15 = 120/240 = 0.5$
$P(DIFFERENT) = 6/16 \times 10/15 + 10/16 \times 6/15 = 120/240 = 0.5$

Page 32 (Extension). This calculator program simulates versions 1 and 2 of game 2. Compare the empirical results with those from the mathematical model.

Note: At this level, we are assuming random drawings at each stage with all choices having equal probabilities. However, suppose the long-term playing strategy does not have to remain fixed and the probabilities of the different choices can vary. Then in some situations such as game 2 (version 2), fair-game playing strategies are possible. Just balance choices equally between even and odd.

Choose 1 with probability 1/2 and 2 with probability 1/2.
If you play this strategy and your opponent selects randomly, the probability for winning is 1/2.
$$1/2 \times 5/9 + 1/2 \times 4/9 = 9/18 = 0.5$$
If you and your opponent both play this strategy, the probability for winning remains 1/2.
$$1/2 \times 1/2 + 1/2 \times 1/2 = 2/4 = 0.5$$

Here are other examples of this generalized strategy of choosing even half the time and odd the other half:

Choose 2 with probability 1/2 and 1 and 3 each with probability 1/4.
Choose 2 with probability 1/2, 1 with probability 1/3, and 3 with probability 1/6.

Extension—GAME SIMULATION ON A PROGRAMMABLE CALCULATOR

Two players, EVEN and ODD, each randomly choose a number from 0 through 9. If the sum of the two numbers is even, player EVEN wins. If the sum of the two numbers is odd, player ODD wins.

In this program for the CASIO 7700, 200 plays are simulated with the proportion for an even sum displayed cumulatively in intervals of 20 plays. Press EXEC to continue. Minor modifications can be made for use on other calculators.

1. $0 \to E$
2. $0 \to N$
3. Lbl 1
4. $0 \to C$
5. Lbl 2
6. Int (10 Ran #) $\to A$
7. Int (10 Ran #) $\to B$
8. $A+B \to S$
9. $S \div 2 = $ Int $(S \div 2) \Rightarrow E+1 \to E$
10. $N+1 \to N$
11. $C+1 \to C$
12. $C < 20 \Rightarrow$ Goto 2
13. "P(EVEN) ="
14. $E \div N$ ◢
15. $N < 200 \Rightarrow$ Goto 1
16. "START AGAIN" ◢

Run the program, recording cumulative results, in intervals of 20, for 200 plays.

1. Describe your results. Do they support the argument that this is a mathematically fair game, where each player has a probability of one-half of winning?

Suppose the choices in the game were made from the numbers 1 through 9. To simulate this version of the game, insert these additional lines in the program:

After line 6: $A = 0 \Rightarrow$ Goto 2   After line 7: $B = 0 \Rightarrow$ Goto 2

Run the program again with these changes, recording cumulative results.

2. Describe your results. Do they support your answer to the question in 1?

3. Plot your results, connecting successive points.

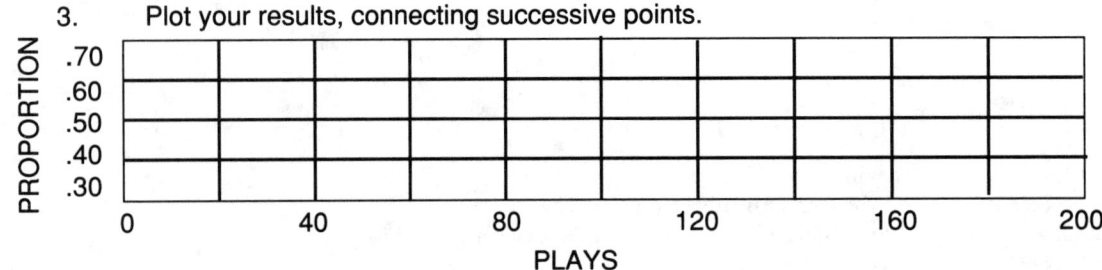

For more long-term results, make these modifications:

Change 20 in line 12 to 200.   Change 200 in line 13 to 2000.

4. Run both versions of the game with these changes. Comment on the results.

Modify the program so that the line curve itself is drawn by a graphing calculator.

Geometry in a Circle

Two friends, Tasha and Hank, were trying to determine where to put a point C on a major arc AB (larger than a semicircle) to form the largest possible angle ACB. Tasha picked C_1; Hank picked C_2.

Who do you think made the better choice? _____

Why? _____

Construct and measure $\angle AC_1B$ and $\angle AC_2B$.
Pick two more points C_3 and C_4 on major arc AB.
Construct and measure $\angle AC_3B$ and $\angle AC_4B$.

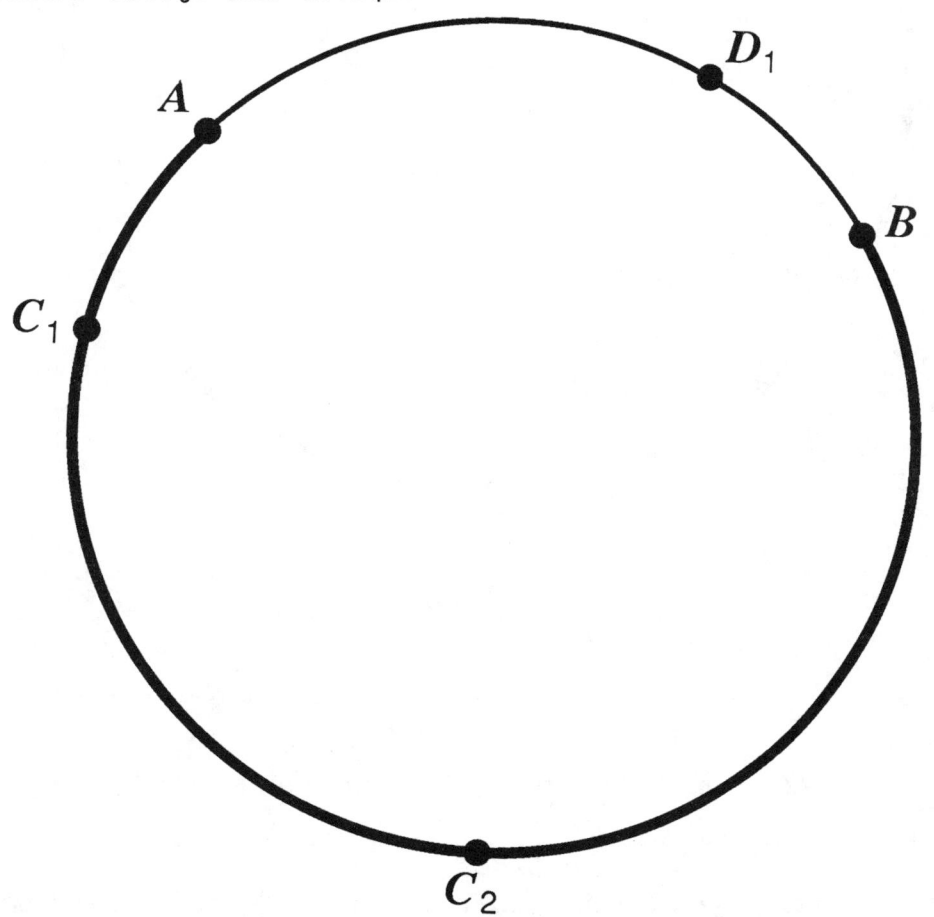

1. $m\angle AC_1B = $ _____, $m\angle AC_2B = $ _____, $m\angle AC_3B = $ _____, $m\angle AC_4B = $ _____
2. What appears to be true about the measures of these angles?

_____

Sherrill thought she could make a larger angle by placing a point D_1 on the minor arc AB (less than a semicircle). Construct and measure $\angle AD_1B$. Pick another point on the minor arc AB and label it D_2. Construct and measure $\angle AD_2B$.

The editors wish to thank Melfried Olson and Judy Olson, Western Illinois University, Macomb, IL 61455, for writing this issue of NCTM Student Math Notes.

3. $m\angle AD_1B = $ _____, $m\angle AD_2B = $ _____

4. What appears to be true about these angles? _____

5. What is the relationship between the measures of $\angle AC_1B$ and $\angle AD_1B$ or $\angle AC_2B$ and $\angle AD_2B$? _____

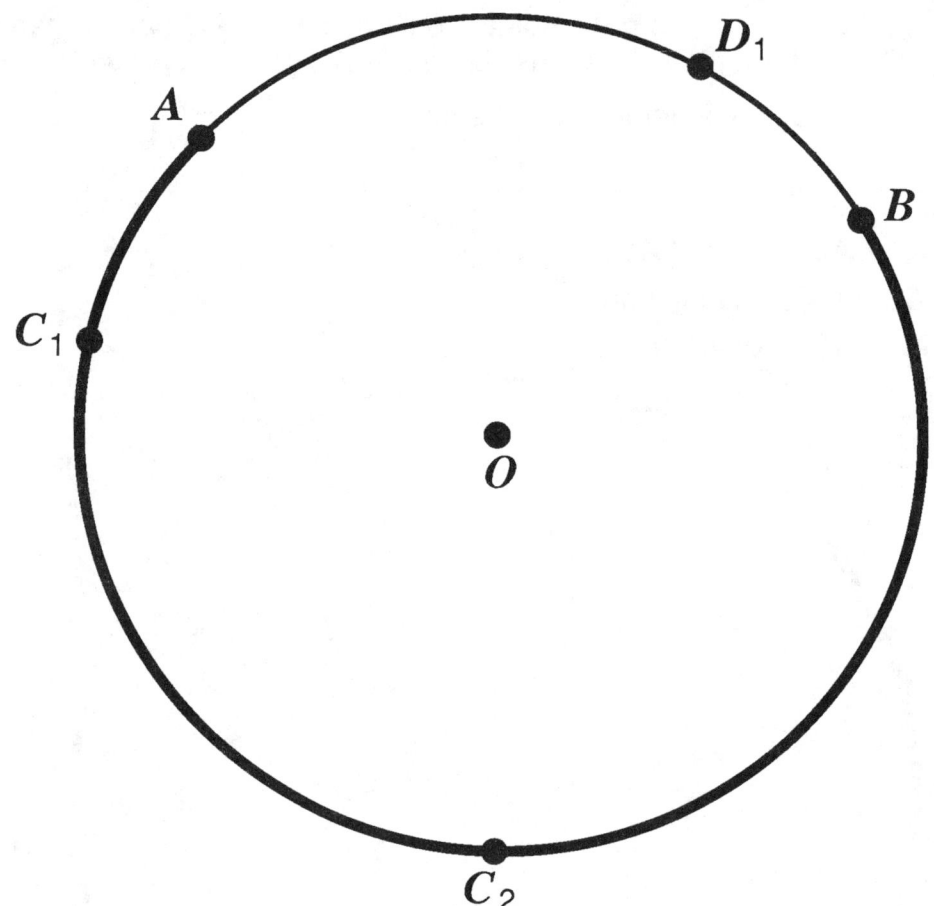

Pete disagreed with the others and thought it might be better to choose the center of the circle, O. Construct and measure $\angle AOB$.

6. $m\angle AOB = $ _____

7. How does $m\angle AC_1B$, $m\angle AC_2B$, $m\angle AC_3B$, or $m\angle AC_4B$ compare with $m\angle AOB$? _____

Draw a line segment from C_1 through O to the other side of the circle. Label the other endpoint F.

8. What is the segment C_1F called? _____

9. What is the arc C_1F called? _____

10. What is the measure of the arc C_1F? _____

Construct and measure $\angle C_1AF$, $\angle C_1D_1F$, and $\angle C_1BF$.

11. $m\angle C_1AF = $ _____, $m\angle C_1D_1F = $ _____, $m\angle C_1BF = $ _____

12. What appears to be true about any angle inscribed in a semicircle? _____

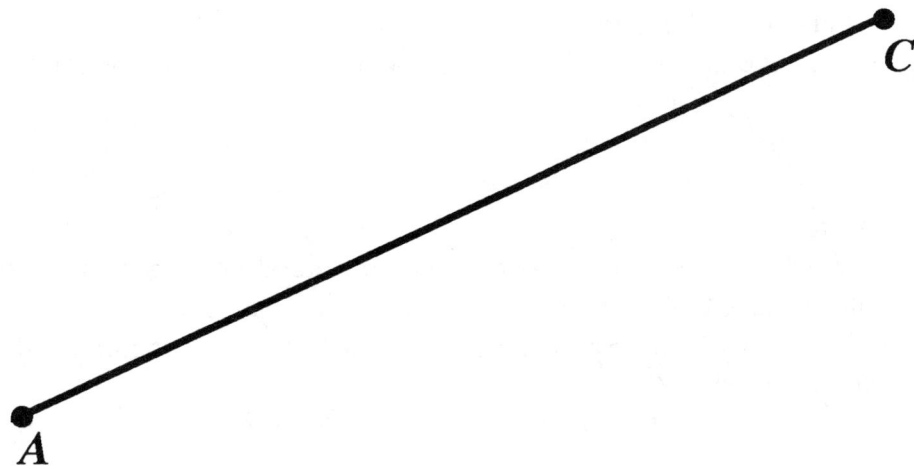

Use a ruler, protractor, or compass to complete the following construction:

- Draw a horizontal line ℓ through A.
- Locate a point P_1 on ℓ to form right triangle AP_1C.
- Draw another line ℓ_1 through A, below $\overline{AC}$, and locate P_2 on ℓ_1 to form right triangle AP_2C.
- Locate several other points P_i below $\overline{AC}$ such that $\triangle AP_iC$ is a right triangle.

13. What figure starts to appear if you look at A, C, and several points P_i? _____

- Locate point O, the midpoint of $\overline{AC}$.
- Draw segment P_1O and extend it to point Q_1 so that $Q_1O = P_1O$.
- Draw other segments P_iO and extend to points Q_i so that $Q_iO = P_iO$.
- Draw $\overline{Q_1A}$, $\overline{Q_1C}$, $\overline{Q_2A}$, $\overline{Q_2C}$, and so on.

14. What kind of polygon is Q_1AP_1C, Q_2AP_2C, and so on? _____

    Why? _____

15. How does AC compare to Q_1P_1, Q_2P_2, and so on?

    _____

16. What part of the rectangle is each of the segments in question 15?

    _____

17. What figure starts to appear if you look at A, C, several points P_i, and several points Q_i? _____

18. The diagonals of the rectangles (for example, Q_1P_1 or AC) are what part of the circle? _____

Did you know that . . .

- any rhombus inscribed in a semicircle is a square?
- the measure of an inscribed angle is one-half the measure of the central angle that intercepts the same arc?
- inscribed angles that intercept the same or congruent arcs are equal in measure?
- a circle passes through the midpoints of the sides of a triangle, the feet of the perpendiculars from the vertices on the sides and the midpoints of the line segments between the vertices and the points of intersection of the altitudes? This circle is known as the nine-point circle.
- travel from a point A to a point B on the earth by plane follows what is known as the great circle route to minimize both fuel and time?

Can you . . .

- inscribe a square in a circle?
- prove that the opposite angles of a cyclic quadrilateral (a quadrilateral inscribed in a circle) are supplementary?
- find $m \angle AD_1 B$ on page 34 if $m \angle AOB = 88$?
- find a strategy to win a contest in which the goal is to construct as many pairs of perpendicular lines as you can in one minute using only a compass and straightedge?
- find the center of this circle using only the corner of a sheet of paper?

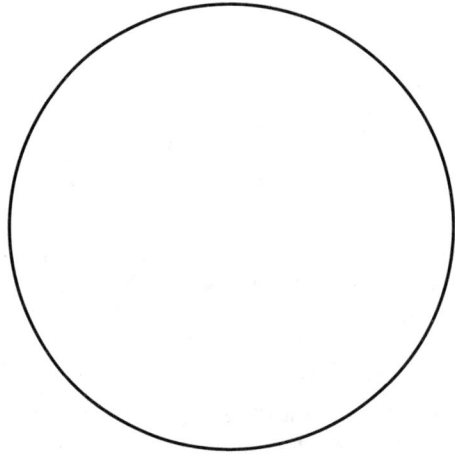

NCTM STUDENT MATH NOTES is published as part of the **NEWS BULLETIN** by the National Council of Teachers of Mathematics, 1906 Association Drive, Reston, VA 22091. The five issues a year appear in September, November, January, March, and May. Pages may be reproduced for classroom use without permission.

Editor:	**Lee E. Yunker,** West Chicago Community High School, West Chicago, IL 60185
Editorial Panel:	**Daniel T. Dolan,** Office of Public Instruction, Helena, MT 59620
	Elizabeth K. Stage, Lawrence Hall of Science, University of California, Berkeley, CA 94720
	John G. Van Beynen, Northern Michigan University, Marquette, MI 49855
Editorial Coordinator:	**Joan Armistead**
Production Assistants:	**Ann M. Butterfield, Karen K. Aiken**

Printed in U.S.A.

Teacher Notes

Page 33. Encourage students to make visual estimates of the measures of angles AC_1B and AC_2B before drawing any segments or measuring any angles. Use a manipulative to support the results found by measuring the various inscribed angles for both the major and minor arcs. Have students cut paper to these angles and see how the vertices can move any place along each arc and still have points A and B on their sides.

As point C moves along the major arc, observe how triangle ABC changes. Have students describe its changing shape in terms of both side and angle classification.

Page 34. Again, use a corner of a sheet of paper and move it so that the endpoints of diameter CF are on two adjacent sides. However the paper is positioned, the corner point, or vertex, always falls on the semicircle formed by the diameter.

Page 35. This activity offers another view of the inscribed right-angle property for a semicircle. But this time, the eye first sees the various rectangles formed with diagonal AC. As point P moves, notice how rectangle $APCQ$ changes. Have your students explore both the changing areas and perimeters of these rectangles, identifying positions for maximum and minimum values.

Page 36. The last question illustrates a hands-on activity connected directly to the right-angle property just established. It offers an excellent cooperative problem-solving experience for small groups of students and a chance for verbal communication, observations, and arguments.

Answers

Page 33. At best, even accurate measurements with a protractor are approximate.
1. The measure of each angle is 52 degrees.
2. All angles have the same measure in degrees.

Page 34.
3. The measure of each angle is 128 degrees.
4. All angles have the same measure in degrees.
5. The sum of the measures of the two angles is 180 degrees. The two angles are supplementary.
6. $m \angle AOB = 104$.
7. Measures of angles AC_1B, AC_2B, AC_3B, and AC_4B are all equal to half that of angle AOB.
8. Diameter
9. Semicircle
10. 180
11. The measure of each angle is 90 degrees.
12. Right angle

Page 35.
13. Semicircle
14. Rectangles; answers will vary.
15. Equal
16. Diagonal
17. Circle
18. Diameter

Page 36. The measure of angle AD_1B is 136.

If $m \angle AOB = 88$ in the figure on page 34, then $m \angle AC_2B = 44$. It follows that $m \angle AD_1B = 136$, since the angles in the major and minor arcs formed by A and B must be supplementary.

Page 38 (Extension). In each case, the vertices P, Q, and R must lie on a pair of congruent major arcs of circles, one above and the other below the chord AB. These arcs locate vertices of inscribed angles of a given size. As the measure of the angle increases, the diameter of the circles forming these arcs decreases until the inscribed angle is 90°, the chord AB is a diameter, and the arcs are semicircles.

Extension—GEOMETRY ON A CIRCLE

All points on the major arc AB containing point C form inscribed angles of 50°, the same size as angle ACB. All points on the minor arc AB containing point D form inscribed angles of 130°, the same size as angle ADC. But where are all the points P that form, with A and B, angles of other sizes, such as 40°?

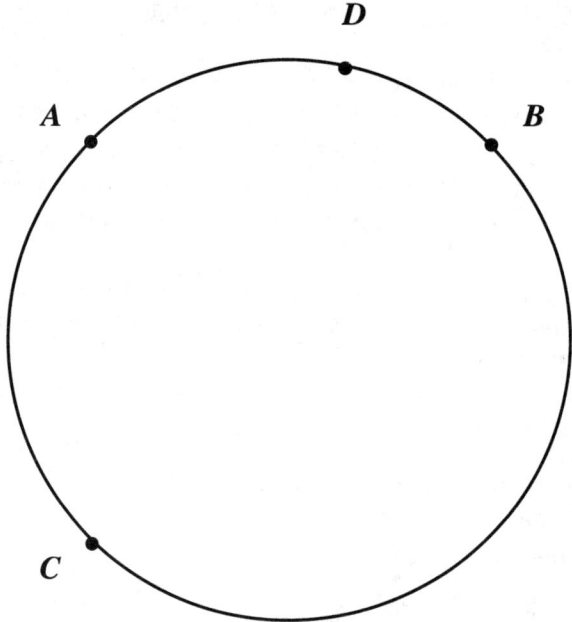

Cut a piece of paper to form a 40° angle. Call the vertex P. Move it between points A and B so that P is at C. Then move it down further until points A and B on the circle touch the sides of the angle. Mark the location of vertex P on the paper using P_1. Move the vertex of the paper angle to another position, still keeping points A and B on the sides. Mark enough different possible locations of vertex P, using P_2, P_3, P_4, and so on, until you can safely guess where points P_i must lie.

1. If angle APB is a 40° angle, on what figure must all the points P lie?

2. Where do you think point Q must lie if angle AQB is 30°?

3. What about the locations, or possible locations, of vertex R of angle ARB if its measurement is 60°? 70°? 80°?

4. Where must vertex S of angle ASB lie if its measure is 90°?

Paper-punching Patterns

Follow the directions and discover some interesting numerical patterns.

- Punch a hole in a sheet of paper.
- Fold the paper in half and punch through the doubled sheet. (Make sure the second punch is different from the first.)
- Guess the number of holes in the paper if it is unfolded.
- Unfold the paper, check your guess, and record your results in table 1.
- Refold the paper as it was; fold it in half once more and punch through all the layers.
- Guess the total number of holes, check your guess, and record the results.
- Record the number of sections formed.
- Use the pattern you observe to extend the pattern beyond the values that you can actually punch.

Do your results look like these?

 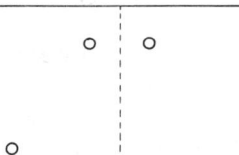 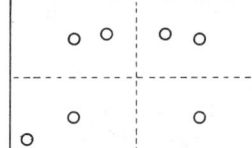 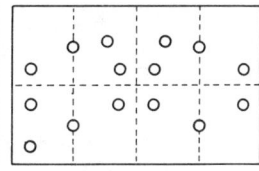

Table 1

Number of times folded in half	Number of sections	Number of holes
0		
1		
2		
3		
4		
5		
6		
7		
8		

What do you think will happen if you vary the number of sections or punches on each turn? You'll have a chance to explore these questions in the following examples.

First, let's try changing the way we fold the paper:

- Punch a hole in a new sheet of paper.
- Fold the paper in thirds and punch all layers as before.
- Fold the paper in thirds a second time and punch through all the layers.
- Guess the total number of sections formed in the paper after each folding and guess the number of holes.
- Unfold the paper and record your results in table 2.
- Extend the pattern in table 2 to find values you can no longer fold or punch.

Do your results look like these?

 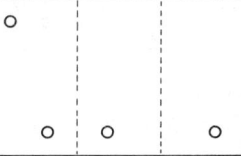 

Table 2

Number of times folded in thirds	Number of sections	Number of holes
0		
1		
2		
3		
4		
5		
6		
7		
8		

Do you see a relationship between the first and second columns of table 1 and table 2? Do you see a relationship between the second and third columns of table 1 and table 2? Are the patterns the same?

The editors wish to thank John Firkins, Gonzaga University, Spokane, WA 99258, for writing this issue of *NCTM Student Math Notes*.

Extending the Patterns

Data from tables 1 and 2 are recorded in a different way below to make one of the patterns clearer. Can you complete tables 3 and 4 for any whole number n?

How are the nth rows of tables 3 and 4 similar to each other? _____

How are they different? _____

Data from table 3 have been recorded in another way in table 5 to make a different pattern clearer. Can you complete this table for any n?

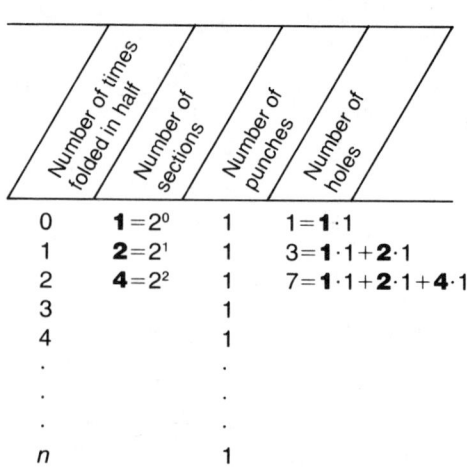

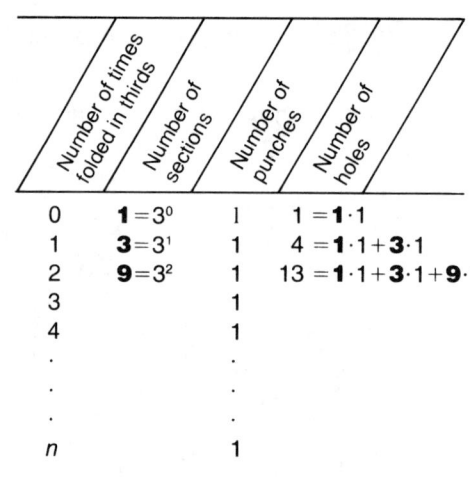

Using powers of 3, see if you can develop a similar rule for table 4.

Suppose you fold a piece of paper into fourths or fifths and punch a hole each time you complete the folding. What would the results be? Complete tables 6 and 7.

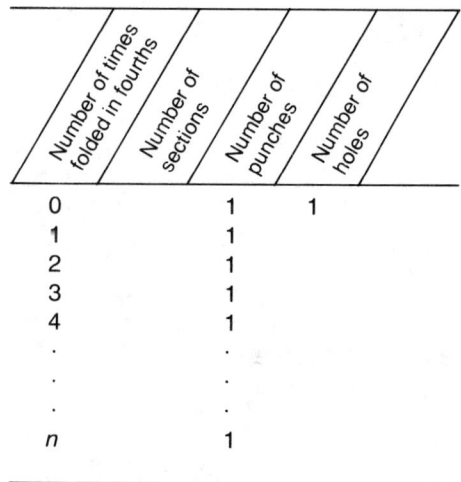

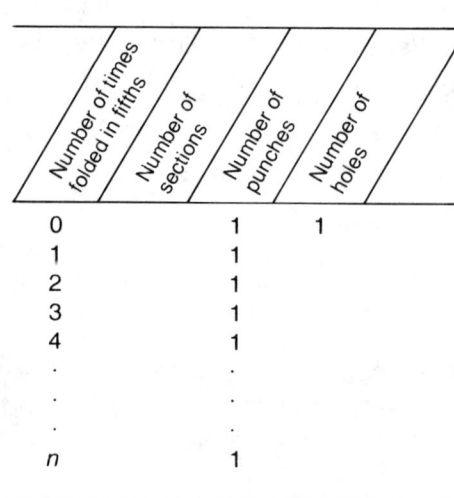

More Patterns

Now we're going to change the number of punches:

- Punch a hole in a new sheet of paper.
- Fold it in half and punch it two more times.
- Guess how many holes you will see when you unfold the paper.
- Unfold the paper and record your findings in table 8.
- Refold the paper, fold it in half again, and punch three times through the paper; record the results.
- Extend the pattern.

Table 8

Number of times folded in half	Number of sections	Number of punches	Number of holes
0	**1**	1	$1 = \mathbf{1} \cdot 1$
1	**2**	2	$5 = \mathbf{1} \cdot 1 + \mathbf{2} \cdot 2$
2	**4**	3	$17 = \mathbf{1} \cdot 1 + \mathbf{2} \cdot 2 + \mathbf{4} \cdot 3$
3			
4			
5			
.			
n			

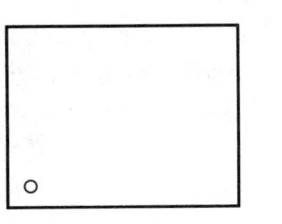
1 hole

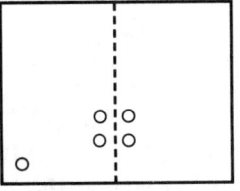
5 holes

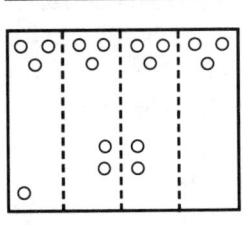
17 holes

Make up your own rule for folding and punching: _____

Using your rule, punch holes in a new sheet of paper and record your results in table 9.

Table 9

Number of times folded	Number of sections	Number of punches	Number of holes
0			
1			
2			
3			
4			
5			
.			
n			

Teaching with *Student Math Notes:* Volume 2

Can you . . .

- fold a square sheet of paper and make one punch so that you will have each of the patterns at the right when you open it? (Use a separate piece of paper for each pattern.)

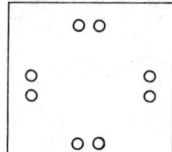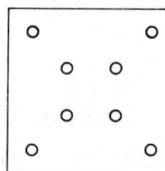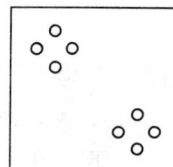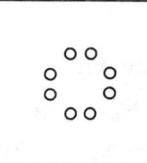

- recognize this pattern? Punch a hole in a new piece of paper. Fold it in half and punch through the same hole. Now fold it in thirds and again punch through the same hole. Continue to extend this pattern in your "mind's eye" by folding the paper in fourths, fifths, sixths, and so on, always punching through the same hole. What familiar number pattern is generated?
- describe a hole-punching activity to obtain the sequence 1, 5, 23, 119, . . . ?
- continue the pattern of punching and folding as seen in tables 5, 6, and 7 from fifths to sixths to sevenths and so on? What do you think will be the general formula for the number of holes after n foldings of the paper into $k$ths?

Did you know . . .

- that the sum of the divisors of some numbers can be determined by hole punching? The sum of the divisors of 8 is 15. The first hole-punching activity produced the same result. Write 8 as 2 to the third, the third power of the prime 2. The prime, 2, tells us to fold the paper in half at each step, and the power of the prime, 3, tells us to continue the hole-punching activity until we have folded the paper in half for the third time. Thus, if a number is a power of a prime, the foregoing activities calculate the sum of its divisors. The formula for computing the sum of the divisors of a power of a prime is given by

$$S = \frac{p^{k+1} - 1}{p - 1},$$

where p is the prime number and k is the power of the prime.

Teacher Notes

Page 39. These hands-on activities connect visualization and geometry to number patterns and problem solving. Have students carefully follow the *fold, punch, guess, unfold* sequence of steps and then count to check. The guessing step is an important part of the process and should not be omitted.

Folding in half can be done in several ways other than the one shown. All folds could be vertical or all horizontal. As long as the folded halves coincide, the results will remain the same.

Page 40. The emphasis here is on extending number patterns. You may want students to actually fold and punch some of the initial stages before completing tables 6 and 7.

Page 41. More number patterns occur when an increasing number of punches is made on successive folds. Encourage students to be creative in their folding and punching and in exploring the resulting number patterns when completing their own versions in table 9.

Answers

Page 39.

Table 1		
0	1	1
1	2	3
2	4	7
3	8	15
4	16	31
5	32	63
6	64	127
7	128	255
8	256	511

Table 2		
0	1	1
1	3	4
2	9	13
3	27	90
4	81	121
5	243	364
6	729	1093
7	2187	3280
8	6561	9841

Page 40.
- Table 3: For n folds, 2^n sections and $1 \times 1 + 2 \times 1 + 4 \times 1 + 8 \times 1 + \cdots + 2^n \times 1$ holes
- Table 4: For n folds, 3^n sections and $1 \times 1 + 3 \times 1 + 9 \times 1 + 27 \times 1 + \cdots + 3^n \times 1$ holes
- Table 5: For n folds, $2^{n+1} - 1$ holes when folded in half
  For n folds, $(3^{n+1} - 1)/2$ holes when folded in thirds

Table 6		
0	1	1
1	4	5
2	16	21
3	64	85
4	256	341
...	...	...
n	4^n	$(4^{n+1} - 1)/3$

Table 7		
0	1	1
1	5	6
2	25	31
3	125	156
4	625	781
...	...	...
n	$5n$	$(5^{n+1} - 1)/4$

Page 41.

Table 8			
0	1	1	1
1	2	2	5
2	4	3	17
3	8	4	49
4	16	5	129
5	32	6	321
...	...	...	...
n	2^n	$n + 1$	$n(2^{n+1}) + 1$

Page 42.

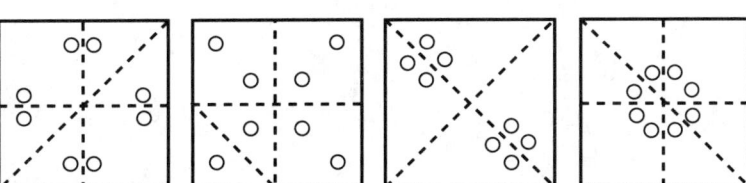

- $1 = 1 = 1!$
  $2 = 1 \times 2 = 2!$
  $6 = 1 \times 2 \times 3 = 3!$
  $24 = 1 \times 2 \times 3 \times 4 = 4!$
  $120 = 1 \times 2 \times 3 \times 4 \times 5 = 5!$ For n parts, $1 \times 2 \times 3 \times 4 \times 5 \times \cdots \times n = n!$

- Punch a hole in a new sheet of paper. Fold it in half and punch it twice, fold it in thirds and punch it three times, and so on.
  $1 = 1 \times 1$
  $5 = 1 \times 1 + 1 \times 4$
  $23 = 1 \times 1 + 1 \times 4 + 2 \times 9$
  $119 = 1 \times 1 + 1 \times 4 + 2 \times 9 + 6 \times 16$ $(k^{n+1} - 1)/(k - 1)$ holes after n foldings of the paper into $k$ths

Extension—PREDICTING PUNCHED PATTERNS

Square sheets of paper are folded and punched as shown. Draw in the folds and the holes when opened.

                Fold 1   Fold 2   Fold 3   Open

NATIONAL COUNCIL OF TEACHERS OF MATHEMATICS   JANUARY 1988

Midpoint Madness

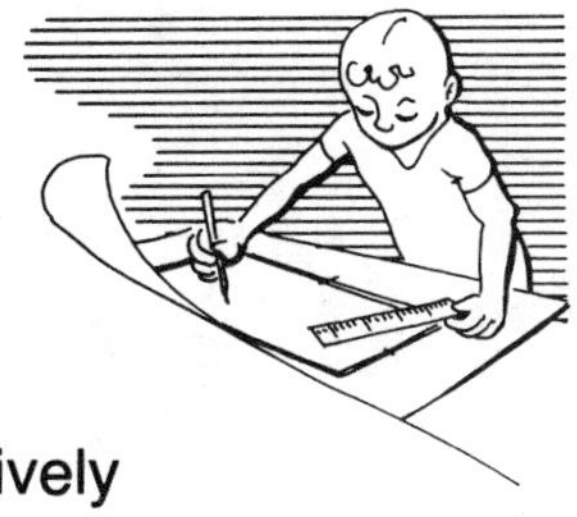

The concept of midpoint is very rich in terms of its usefulness in mathematics. In each of the following quadrilaterals find the midpoints of the sides and label them consecutively *N*, *C*, *T*, and *M*. Then draw the segments forming the polygon *NCTM*.

Square

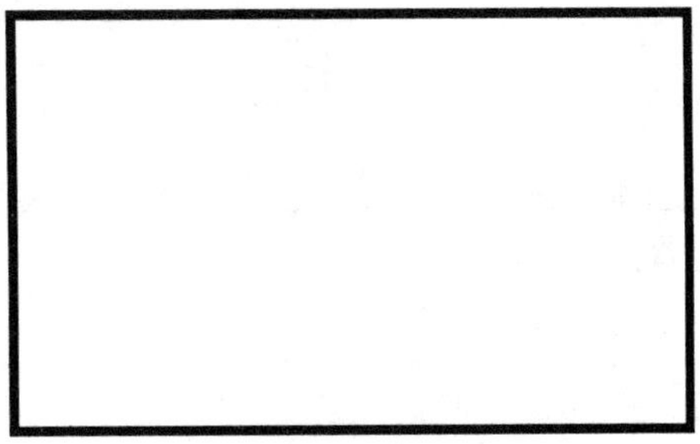

Rectangle

Parallelogram

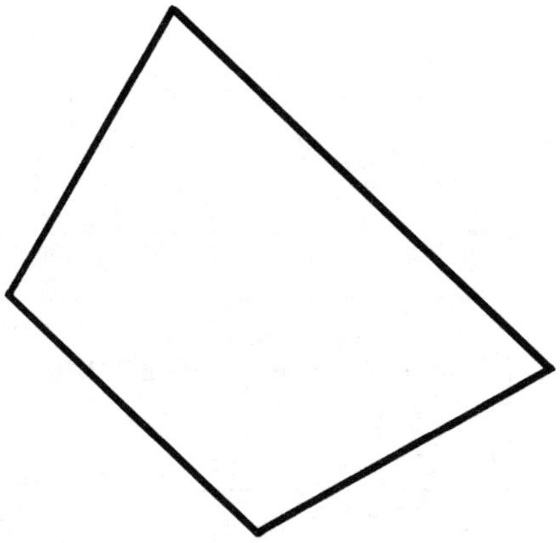

Trapezoid

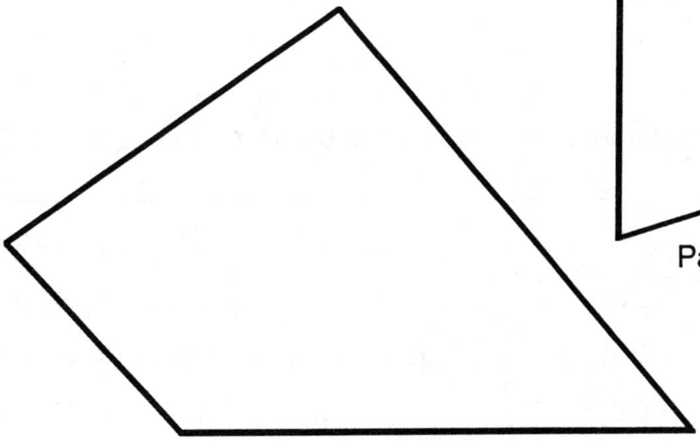

General quadrilateral

What appears to be true about all the *NCTM* polygons? _____

What also appears to be true about the relationship between the area of each *NCTM* polygon and the area of its original quadrilateral? _____

For centuries mathematics students have studied a theorem stating that connecting the midpoints of consecutive sides of any quadrilateral will always form a parallelogram and that the area of the parallelogram is one-half the area of the original quadrilateral.

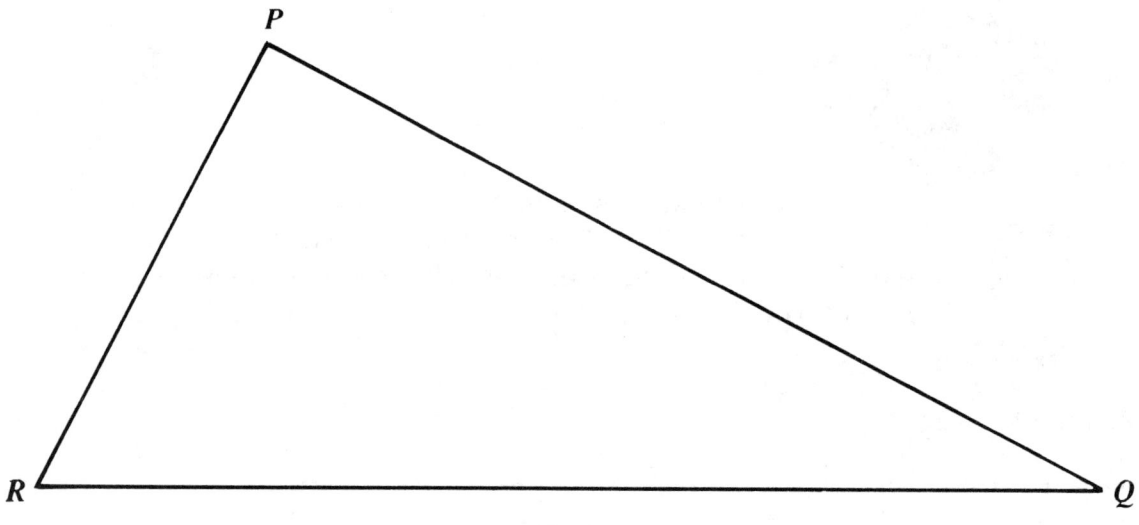

Fig. 1

In $\triangle PQR$, locate the midpoints of sides $\overline{PQ}$, $\overline{QR}$, and $\overline{RP}$ and label them A_1, B_1, and C_1, respectively. Connect these midpoints, forming $\triangle A_1B_1C_1$.

- How do $m\angle P$ and $m\angle A_1B_1C_1$ compare? _____
- How do $m\angle Q$ and $m\angle A_1C_1B_1$ compare? _____
- How do $\triangle PQR$ and $\triangle B_1C_1A_1$ compare in terms of shape? _____
- How do they compare in terms of area? _____
- What kinds of quadrilaterals are polygons $A_1B_1C_1P$, $A_1QB_1C_1$, and $A_1B_1RC_1$?

  _____

  Why? _____

- What kinds of quadrilaterals are polygons A_1B_1RP, C_1A_1QR, and B_1C_1PQ?

  _____

  Why? _____

Find the midpoints of the sides of $\triangle A_1B_1C_1$ and name them A_2, B_2, and C_2. (You may label them in any order.)

- How do $\triangle PQR$, $\triangle A_1B_1C_1$, and $\triangle A_2B_2C_2$ compare in terms of shapes?

  _____

  How do they compare in terms of areas? _____

Repeat the process of successively finding midpoints of the sides of the newly formed interior triangles, labeling them A_i, B_i, and C_i as the number of triangles increases from 1 to 2 to 3 to i.

- What do you notice about the points A_i, B_i, and C_i as i increases?

  _____

Draw in the medians $\overline{RA_1}$, $\overline{PB_1}$, and $\overline{QC_1}$. Label their point of intersection O.

- What do you notice about O and the points A_i, B_i, and C_i as i increases?

  _____

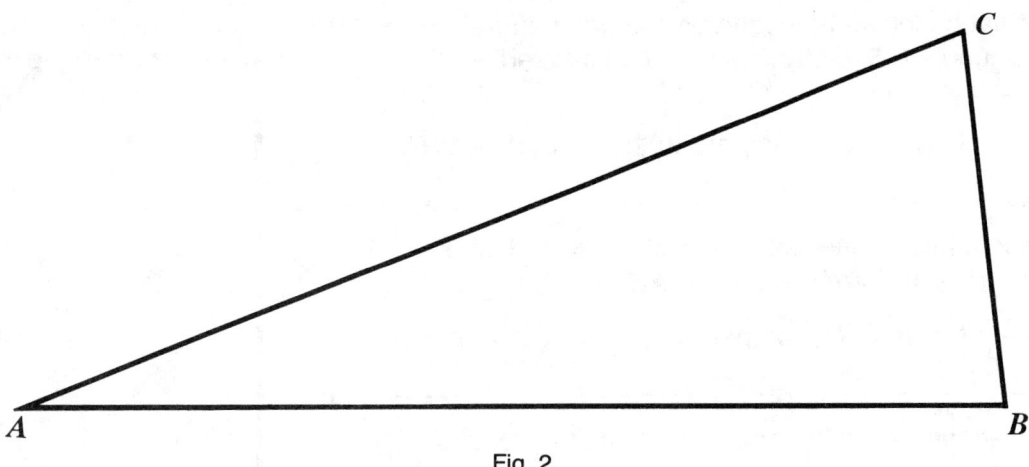

Fig. 2

In $\triangle ABC$, locate D_1, the midpoint of $\overline{AC}$, and E_1, the midpoint of $\overline{BC}$. Draw $\overline{D_1E_1}$; measure $\overline{D_1E_1}$ and $\overline{AB}$.

$D_1E_1 =$ _____ ; $AB =$ _____ .

- How does D_1E_1 compare to AB? _____

Draw $\overline{D_1B}$ and $\overline{E_1A}$; label the point of intersection F_1.

Locate D_2, the midpoint of $\overline{AF_1}$, and E_2, the midpoint of $\overline{BF_1}$. Draw $\overline{D_2E_2}$; measure $\overline{D_2E_2}$.

$D_2E_2 =$ _____ .

- How does D_2E_2 compare to D_1E_1? _____
- How does D_2E_2 compare to AB? _____

Draw $\overline{D_2B}$ and $\overline{E_2A}$; label the point of intersection F_2.

Locate D_3, the midpoint of $\overline{AF_2}$, and E_3, the midpoint of $\overline{BF_2}$. Draw $\overline{D_3E_3}$; measure $\overline{D_3E_3}$. $D_3E_3 =$ _____ .

- How does D_3E_3 compare to D_1E_1 and D_2E_2?

_____

- How does D_3E_3 compare to AB?

_____

- If we continue, how would the measures of all the segments D_iE_i compare to one another?

_____

- If we continue, how would the measures of all the segments D_iE_i compare to the measure of $\overline{AB}$?

_____

- What do you notice about the points F_i as i increases?

_____

Draw $\overline{D_1D_2}$, $\overline{D_2D_3}$, $\overline{E_1E_2}$, $\overline{E_2E_3}$, and so on.

- What kind of quadrilateral is $D_1D_2E_2E_1$ or $D_2D_3E_3E_2$ or $D_1D_3E_3E_1$? _____

  Why? _____

- What kind of quadrilateral is AD_1E_1B or AD_2E_2B or AD_3E_3B? _____

  Why? _____

The midpoints of each side were located in the unit square $A_1B_1C_1D_1$. The square $A_2B_2C_2D_2$ was then constructed.

How does the area of $A_2B_2C_2D_2$ compare to the area of $A_1B_1C_1D_1$?

___

Now locate the midpoints of each side of $A_2B_2C_2D_2$ and label them A_3, B_3, C_3, and D_3 to form the square $A_3B_3C_3D_3$.

How does the area of $A_3B_3C_3D_3$ compare to the area of $A_1B_1C_1D_1$?

___

Continue to generate several more squares using the process described previously. How does the area of $A_5B_5C_5D_5$ compare to the area of $A_1B_1C_1D_1$?

___

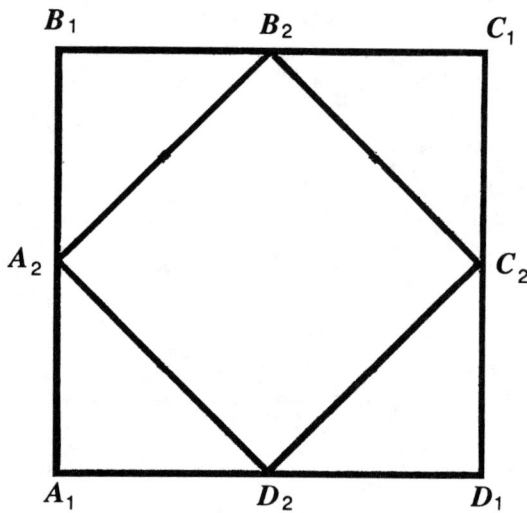

How does the area of $A_9B_9C_9D_9$ compare to the area of $A_1B_1C_1D_1$? ___

What is the area of $A_iB_iC_iD_i$ approaching as i gets larger? ___

With a colored pencil, connect segments A_1A_2, A_2A_3, A_3A_4, and A_4A_5.

1. What is the sum of the measures of these four segments? ___
2. What is the sum of the measures of n segments (as n gets extremely large)? ___
3. Assume that the squares are generated ad infinitum. What is the measure of $A_1A_2 + A_2A_3 + A_3A_4 + \ldots$?

Did you know that . . .

- the point O in figure 1 is called the centroid of triangle PQR? In physics, centroids are known as the center of mass.
- the sequence ½, ¼, ⅛, . . . , $½^n$ is known as a geometric sequence?
- when the terms of a geometric sequence are added together, a geometric series is formed?
- the sum of the geometric series $1 + ½ + ¼ + ⅛ + \ldots$ is 2?

Can you . . .

- prove that the NCTM polygons are all parallelograms?
- prove that the area of each NCTM polygon is half the area of its original quadrilateral?
- prove that the area of $\triangle A_1B_1C_1$ in figure 1 is one-fourth the area of $\triangle PQR$?
- prove that the area of $\triangle CD_1E_1$ in figure 2 is one-fourth the area of $\triangle ABC$?

___

NCTM STUDENT MATH NOTES is published as part of the **NEWS BULLETIN** by the National Council of Teachers of Mathematics, 1906 Association Drive, Reston, VA 22091. The five issues a year appear in September, November, January, March, and May. Pages may be reproduced for classroom use without permission.

Editor: Lee E. Yunker, West Chicago Community High School, West Chicago, IL 60185
Editorial Panel: Daniel T. Dolan, Office of Public Instruction, Helena, MT 59620
Elizabeth K. Stage, Lawrence Hall of Science, University of California, Berkeley, CA 94720
John G. Van Beynen, Northern Michigan University, Marquette, MI 49855
Editorial Coordinator: Joan Armistead
Production Assistants: Ann M. Butterfield, Karen K. Aiken

Printed in U.S.A.

Teacher Notes

Page 45. Students can locate the midpoints by carefully measuring with a ruler. The first figure, formed in a square, is a square, and the second, formed in a rectangle, is a rhombus. But the visual classification that identifies all such midpoint polygons NCTM is *parallelogram*. It is far less apparent, visually, that the new area in the parallelogram is always half that of the original. A vivid demonstration can be made on an overhead projector by cutting off the corner pieces and showing that they can be arranged to form the remaining parallelogram.

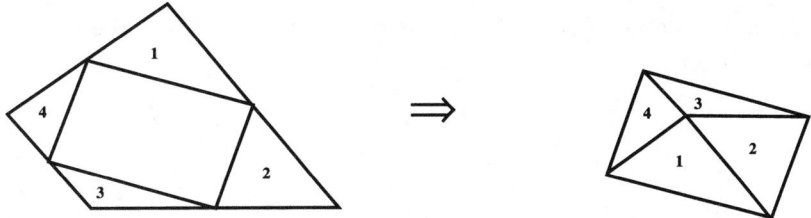

Page 46. Accurate measuring and drawing are essential here. Cutout triangles on an overhead projector again offer convincing proof of both the similarity and area relationships.

Page 47. It is critical that careful drawings be made here in order to establish visually the required properties. A more abstract approach uses the midpoint theorem from the previous page, noting that the base of every triangle AF_iB remains fixed.

Page 48. Paper folding can be a useful activity here. In questions 1–3, do not let students get so involved in the arithmetic and algebra of the geometric sequence that they miss the geometric view of a beautiful spiral turning around toward the center point of the original square.

Answers

Page 45. For each quadrilateral, the polygon NCTM formed by connecting successive midpoints appears—
  a) to be a parallelogram;
  b) to have an area one-half that of the original polygon.

Page 46.
- The measures are equal.
- The measures are equal.
- The triangles are similar.
- The smaller triangle has one-fourth the area of the larger triangle.
- The quadrilaterals are parallelograms because their opposite sides are parallel.
- The quadrilaterals are trapezoids because exactly one pair of opposite sides are parallel.
- All three triangles are similar, and their areas, greatest to least, are in the ratio of 1 to 1/4 to 1/16.
- The three points appear to converge to a single point as i increases without bound.
- The three points appear to converge to the single median intersection point O.

Page 47.
- The shorter segment D_1E_1 (6.5 cm) is half the length of the longer segment AB (13.0 cm).
- The segment D_2E_2 (6.5 cm) has the same length as segment D_1E_1 and half that of segment AB.
- The segment D_3E_3 (6.5 cm) has the same length as segments D_2E_2 and D_1E_1 and half the length of segment AB.
- They would all be equal.
- The ratio of the lengths of D_iE_i to AB will always be 1 to 2.
- They are collinear.
- They are parallelograms because they always have a pair of parallel and congruent sides.
- They are trapezoids because they always have one pair of parallel sides.

Answers, continued

Page 48. Square $A_2B_2C_2D_2$ has 1/2 the area of $A_1B_1C_1D_1$.
Square $A_3B_3C_3D_3$ has $(1/2)^2$, or 1/4, the area of $A_1B_1C_1D_1$.
Square $A_5B_5C_5D_5$ has $(1/2)^4$, or 1/16, the area of square $A_1B_1C_1D_1$, whereas square $A_9B_9C_9D_9$ has $(1/2)^8$, or 1/256, the area. As i gets large without bound, the area of $A_iB_iC_iD_i$ approaches 0.

1. $S = \dfrac{1}{2} + \dfrac{1}{2} \cdot \dfrac{\sqrt{2}}{2} + \dfrac{1}{2} \cdot \dfrac{1}{2} + \dfrac{1}{2} \cdot \dfrac{\sqrt{2}}{4} = \dfrac{1}{2} + \dfrac{\sqrt{2}}{4} + \dfrac{1}{4} + \dfrac{\sqrt{2}}{8} = \dfrac{3}{4} + \dfrac{3\sqrt{2}}{8} = \dfrac{6 + 3\sqrt{2}}{8}.$

2. S is the sum of a finite geometric sequence with the first term $a = 1/2$ and the ratio of successive terms $r = \sqrt{2}/2$. The nth term is
$$\dfrac{1}{2}\left(\dfrac{\sqrt{2}}{2}\right)^{n-1}.$$
Thus
$$S = \dfrac{\dfrac{1}{2} - \dfrac{1}{2}\left(\dfrac{\sqrt{2}}{2}\right)^n}{1 - \dfrac{\sqrt{2}}{2}}.$$

3. S is the sum of an infinite geometric sequence with the first term $a = 1/2$ and the ratio of successive terms $r = \sqrt{2}/2$. Since $|r| < 1$, the series converges to
$$S = \dfrac{\dfrac{1}{2}}{1 - \dfrac{\sqrt{2}}{2}} = \dfrac{2 + \sqrt{2}}{2}.$$

Page 50 (Extension). Triangle A forms a quadrilateral when folded, triangle B forms a triangle, and triangle C forms a concave pentagon. Let P be the folded vertex in triangle PQR, with PQ the shorter of the two sides PQ and PR.

If $PQ = 1/2 \, QR$, then a triangle is formed.
If $PQ > 1/2 \, QR$, then a quadrilateral is formed.
If $PQ < 1/2 \, QR$, then a pentagon is formed.

Extension—MORE MIDPOINT MADNESS

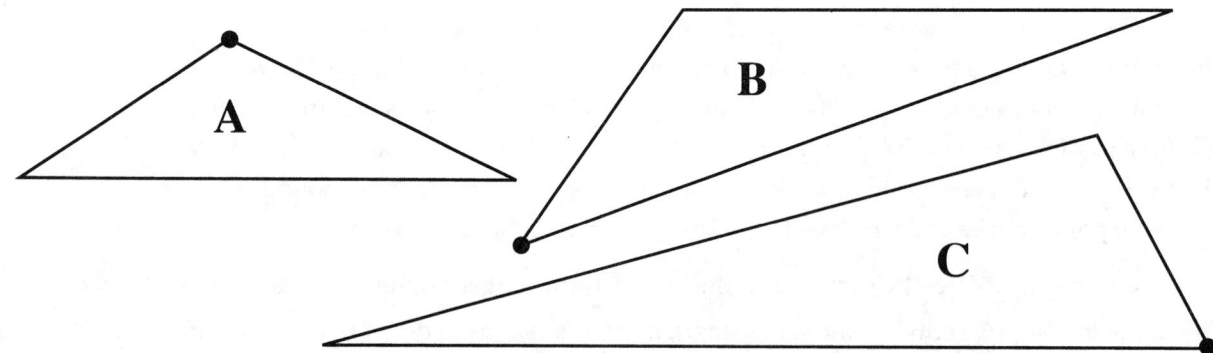

1. Cut out tracings of the three triangles given below. In each case, fold the marked vertex to the midpoint of the opposite side. Name the polygon formed by the boundary of the resulting folded triangle.

2. Cut out another triangle of your own choosing. Select a vertex at random. Try to predict the shape of the polygon that will be formed if that vertex is folded to the midpoint of the opposite side. Then check your answer by folding the paper triangle.

3. See if you can discover what conditions must exist in the original triangle for the folded figure to be a triangle, a

Teaching with *Student Math Notes*: Volume 2

Minimum Distance

Stu Dent recently inherited a large parcel of land shaped like an equilateral triangle. The land's boundaries are three highways, intersecting to form the triangle's three vertices. However, for Stu to receive the land, the will required that he build a house on the triangular parcel of land and also that he build three roads from the house, each perpendicular to one of the three highways. Furthermore, the will required that the sum of the distances, $d_1 + d_2 + d_3$, from the house to the highways *must* be the **smallest** possible. Where should Stu put his house?

Let's Explore Some Possibilities for Stu

Using a ruler and protractor, locate possible points—housing sites—on or within the triangles below and construct the three roads from the house perpendicular to the highways. Measure and record the **sum** of the three distances under each triangle.

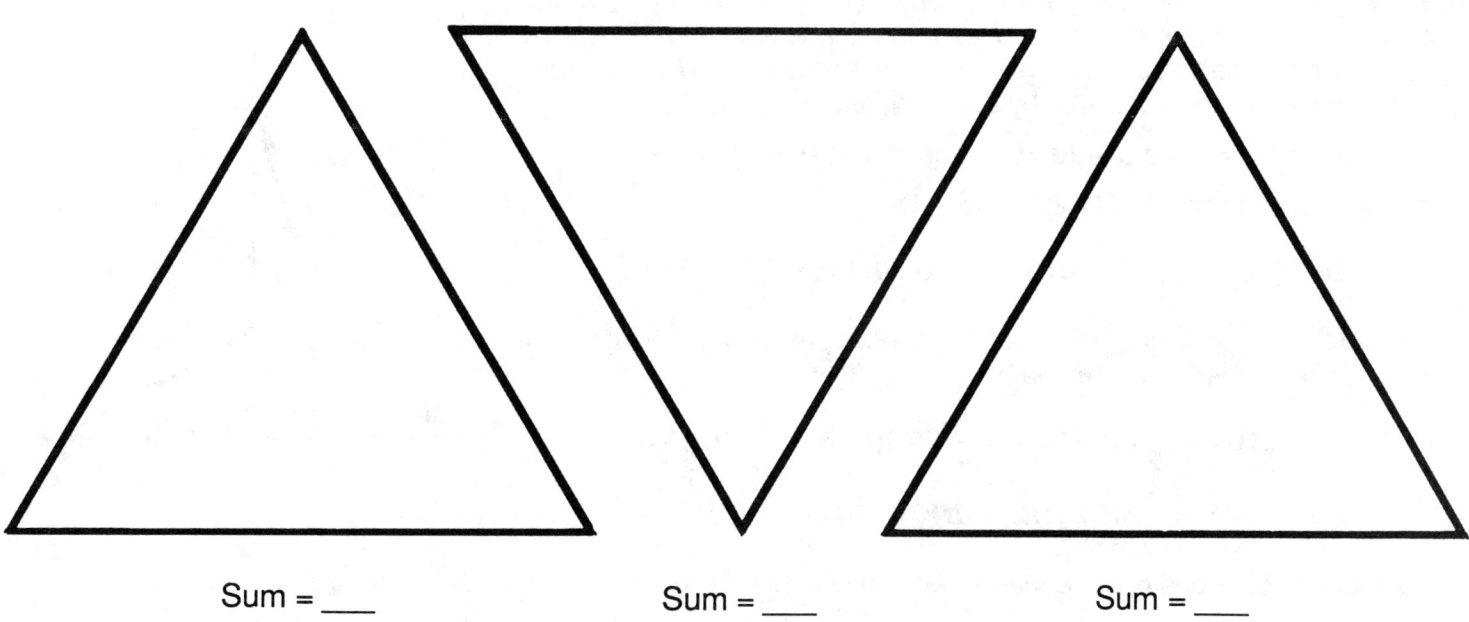

Sum = ___   Sum = ___   Sum = ___

The editors wish to thank Jerry L. Johnson, Western Washington University, Bellingham, WA 98225, for writing this issue of the *NCTM Student Math Notes*.

Continuing Your Explorations

Now try some additional points. Be sure to try special points, such as the intersection of the medians, a vertex, or a point on the edge of the parcel of land.

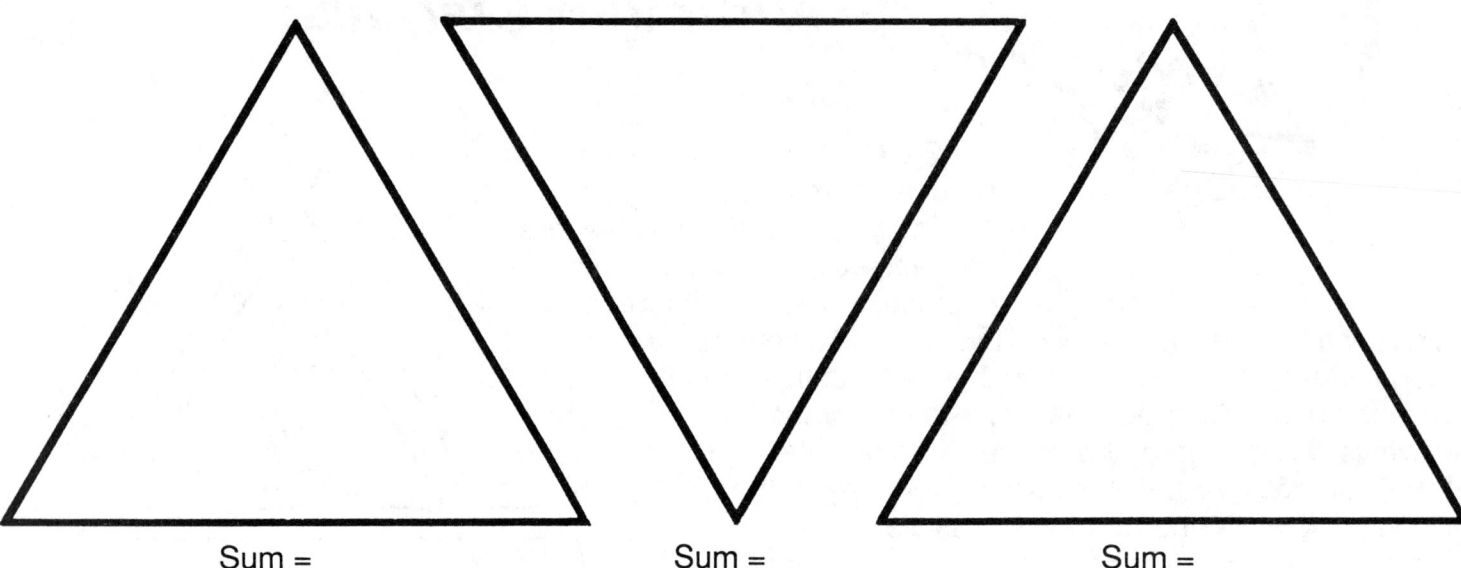

Sum = ___   Sum = ___   Sum = ___

- What is the sum of the distances whenever the point chosen is—

    within the triangle? _____   on the edge of the triangle? _____

    at a vertex of the triangle? _____

- What appears to be true about all these sums? _____

Verifying Your Discovery

The diagram at the right represents the conditions of the will. Point H was selected at random to represent the location of the house with $\overline{HD}$, $\overline{HE}$, and $\overline{HF}$ representing the three roads. Notice that the area of $\triangle ABC$ has been divided into three smaller triangles by $\overline{HC}$, $\overline{HB}$, and $\overline{HA}$, and that

area of $\triangle ABC$ = area of $\triangle BCH$ + area of $\triangle ACH$ + area of $\triangle ABH$.

Using the area formula, $A = (1/2)bh$, we have

$$\tfrac{1}{2}(BC)(AG) = \tfrac{1}{2}(BC)(HE) + \tfrac{1}{2}(AC)(HF) + \tfrac{1}{2}(AB)(HD).$$

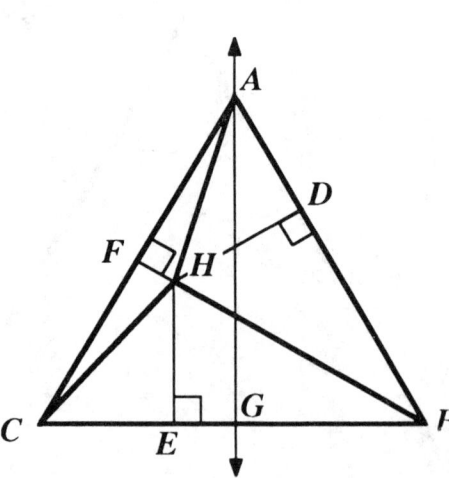

Since $\triangle ABC$ is equilateral, the sides have equal measure. Therefore, by substituting BC for AC and AB, we have

$$\tfrac{1}{2}(BC)(AG) = \tfrac{1}{2}(BC)(HE) + \tfrac{1}{2}(BC)(HF) + \tfrac{1}{2}(BC)(HD)$$

$$= \tfrac{1}{2}(BC)(HE + HF + HD).$$

By comparing both sides of this equation, we can see that $AG = HE + HF + HD$.

Therefore the sum of the three perpendicular segments, or roads, is a constant value. Regardless of where the house is located, the sum of the segments is equal to the length of the altitude of the original equilateral triangle, or land parcel. Now, construct an altitude in one of the triangles at the top of the page and measure its length to verify your discovery. Is its length equal to your sums? _____

Further Explorations

If the parcel of land in the will had been in the shape of a square, where should Stu locate the house for the sum of the distances to the four sides to be a minimum?

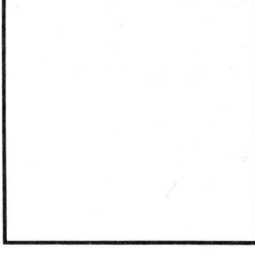 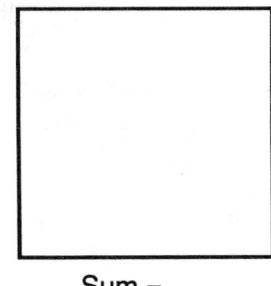 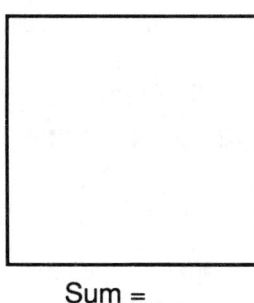

Sum = ___   Sum = ___   Sum = ___

- Where should Stu locate the house? _____
- What relationship, if any, exists between these sums and the dimensions of the square? _____
  _____

If the parcel of land had been a regular hexagon, where should Stu locate the house?

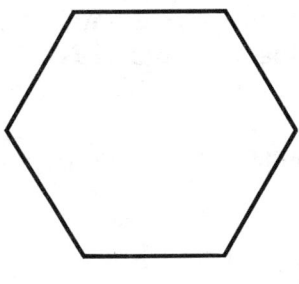 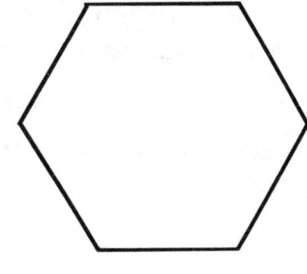 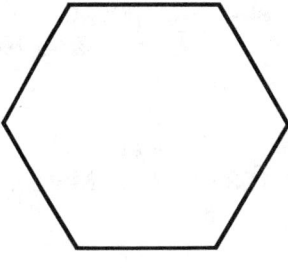

Sum = ___   Sum = ___   Sum = ___

- Where should Stu locate the house? _____
- What relationship, if any, exists between these sums and the dimensions of the regular hexagon? _____
  _____

Exploring Other Shapes

Trace copies of the following figures and continue to explore possible housing sites for each figure that would produce a minimum sum for the perpendicular distances from the house to each side.

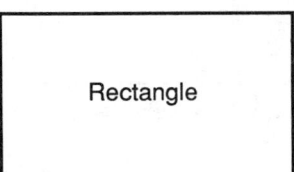 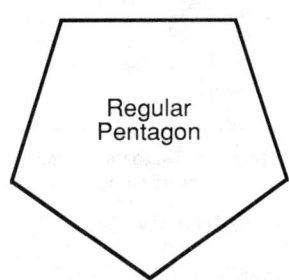 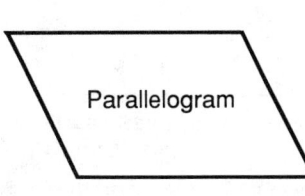 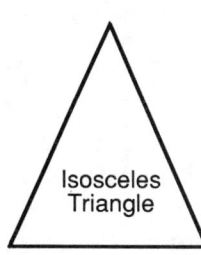

Rectangle   Regular Pentagon   Parallelogram   Isosceles Triangle

- What conclusions can you make regarding the best location for Stu's house in these new shapes? _____

Teaching with *Student Math Notes:* Volume 2

Can you . . .

- prove your discoveries for the **regular** figures other than the equilateral triangle?
- find the best location for Stu's house if the distance to the three vertices of the triangle from the house is to be a minimum?
- write a Logo program that will simulate the original problem visually, calculating the respective lengths and sum as well?
- use the Geometric Supposer or similar software to investigate the original problem?
- write a computer program that uses a coordinate approach to solve the original problem?
- extend the original problem to a regular tetrahedron and other regular polyhedra?

Did you know that . . .

- the study of soap bubbles led to the discovery of a solution to several of the minimum-distance problems in mathematics?
- the original problem involving the equilateral triangle was first solved by the Italian mathematician Vincenzo Viviani (1622–1703)? In fact, Viviani showed that the house could be anywhere (inside or outside the triangle) if the perpendicular distances of a point outside the triangle to the triangle's sides are considered to be negative.
- the French mathematician Pierre de Fermat (1601–1665) posed a related problem solved by the Italian physicist Evangelista Torricelli (1608–1647): Locate a point the sum of whose distances from the vertices of **any** given triangle is a minimum?
- both Viviani and Torricelli were students of Galileo Galilei (1564–1642)?
- in 1935, Paul Erdos posed an interesting conjecture related to the original problem: If H is any point within any $\triangle ABC$, and if $\overline{HD}$, $\overline{HE}$, and $\overline{HF}$ are the perpendicular distances from H to the sides of $\triangle ABC$, then

$$HA + HB + HC \geq 2(HD + HE + HF)?$$

The equality holds if and only if $\triangle ABC$ is equilateral and the point H is its circumcenter. This conjecture was first proved in 1937 by L. J. Mordell and D. R. Barrow.

Teacher Notes

Page 51. Students may need to be reminded that distances from points to segments must be measured in a perpendicular fashion. In this activity, the three perpendicular distances d_1, d_2, and d_3 from each point need to be drawn and measured carefully.

Page 52. It is not intuitively obvious that the sum of the distances from the three sides remains constant for any point on or within the equilateral triangle. Contrast this result with points outside the triangle where the sum is always greater.

When discussing the proof at the bottom of the page, emphasize that H can be located anywhere in the triangle without changing the argument. Encourage students to see how the figure changes as H moves around. A point of particular interest is where H is at the intersection of the three altitudes. One quickly sees that this point must necessarily be equidistant from the three sides and, therefore, two-thirds of the way down each altitude from the corresponding vertex.

Page 53. These questions offer students an excellent opportunity to do their own exploration of problem solving. Some may require larger figures to do accurate drawing and measuring.

Answers

Page 51. Small errors may occur in measuring, but all answers should be approximately 6.7 cm.

Page 52. Whether within, on the edge, or at a vertex, the sum of the three distances should be approximately 6.7 cm. This is the length of each altitude. For an equilateral triangle with sides of 1 each, the altitudes measure $\sqrt{3}/2$.

Page 53. SQUARES: Every point on or within the square has the same constant for the sum of the distances from the four sides. For the square, this sum is always twice the length of a side.

REGULAR HEXAGONS: Every point on or within the regular hexagon has the same constant for the sum of its distances to the six sides. For the regular hexagon, this sum is always $3\sqrt{3}$ times the length of a side. This is six times the distance from the center to one side of the hexagon.

OTHER SHAPES: Every point produces a constant sum for a rectangle, a regular pentagon, or any other regular polygon. However, no obvious or easy answer exists for the other shapes.

Page 56 (Extension). The point P where the three altitudes intersect is the closest to all three vertices. The sum of the distances to the three vertices is at a minimum. The sum is greater for every other point on or within the equilateral triangle. The greatest maximum sum occurs at each of the three vertices A, B, and C.

If triangle DEF has an angle of 120 degrees or more, choose that vertex for the minimum sum. Otherwise, choose the point P within the triangle so that segments PD, PE, and PF form three 120-degree angles for the minimum sum of the distances to the three vertices.

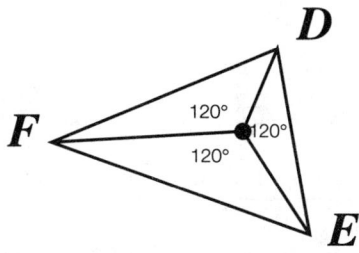

Teaching with *Student Math Notes*: Volume 2

Extension—MORE MINIMUM AND MAXIMUM DISTANCES

Take any point P on or within the equilateral triangle ABC. Regardless of where you choose the point P, the sum of the distances from the three sides remains constant. But suppose you measure the distances from P to the three vertices. At what locations will the sum of these distances be a minimum and a maximum, or will the sum always be the same?

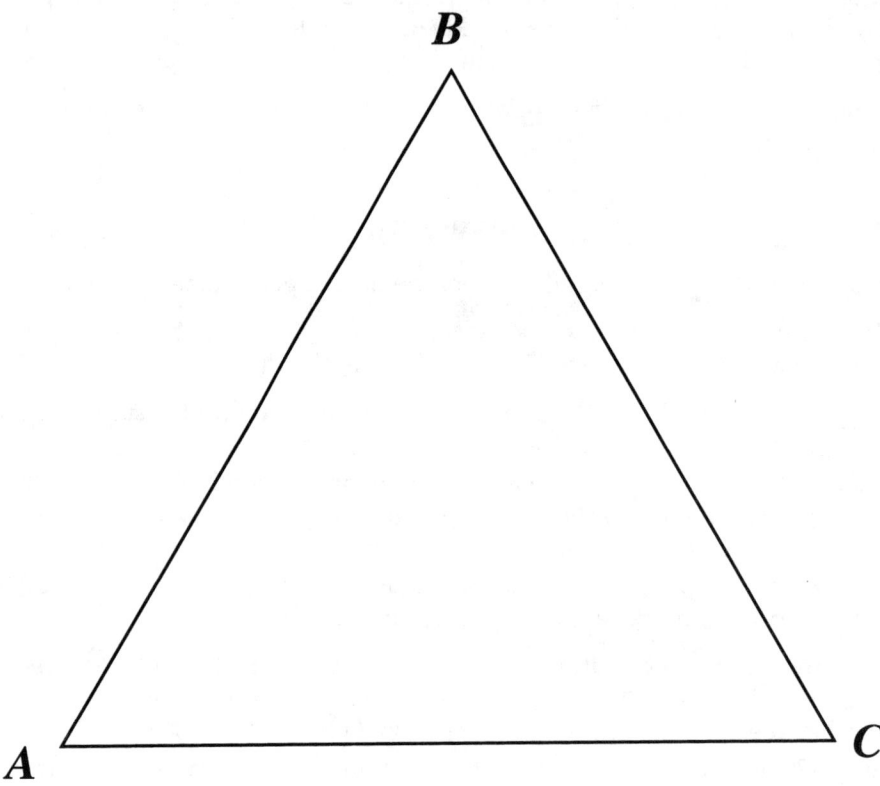

- Select three points, P_1, P_2, and P_3, on or within triangle ABC. Find the sum of the distances to the vertices for each point.

    P_1 _____ P_2 _____ P_3 _____

- Choose some other points. See if you can find where the minumum and maximum sums of distances occur.

- What happens in a triangle DEF that is not equilateral?

Statistical Decision Making

A panel of five experts rated the centers in the National Basketball Association (NBA) on five characteristics: aggressiveness, shooting range, teamwork, offense, and defense. The players were rated on each characteristic from 10 for excellent to 1 for poor. Table 1 gives the totals in each category for ten of the players.

Table 1

Player	Aggressiveness	Ranking	Shooting Range	Ranking	Teamwork	Ranking	Offense	Ranking	Defense	Ranking	Total of Rankings	Order
Brad Daugherty (Cleveland Cavaliers)	41		37		34	5	38		41			
Patrick Ewing (New York Knicks)	47		34		36	2	45		39			
Artis Gilmore (Chicago Bulls)	41		34		32	8	32		33			
Kareem Abdul Jabbar (Los Angeles Lakers)	33		38		48	1	48		35			
Bill Laimbeer (Detroit Pistons)	45		42		35	3	36		31			
Moses Malone (Washington Bullets)	49		37		27	10	46		36			
Akeem Olajuwon (Houston Rockets)	48		36		33	7	48		50			
Robert Parrish (Boston Celtics)	36		28		35	3	39		40			
Tree Rollins (Atlanta Hawks)	46		26		31	9	33		38			
Jack Sikma (Milwaukee Bucks)	33		43		34	5	36		28			

Using these results, who would you choose as the number-one center in the NBA? _____

- A way to find the number-one center is to rank the players in each category from 1 (first) to 10 (last). The rank for teamwork would be 1—Jabbar (48); 2—Ewing (36); 3—Laimbeer, Parrish (35); 5—Daugherty, Sikma (34); 7—Olajuwon (33); 8—Gilmore (32); 9—Rollins (31); 10—Malone (27). Note that no 4 is listed because two players tie for number 3. Rank the other categories, total the ranks for each player, and order the players.

- Another way to select the best center is to determine who has the highest point total from the five characteristics. Find this total for each player and enter the results under "Total Points." Use those results to rank the players in the next column.

- Some people prefer to *weight* or count some of the categories more than others because they think that teamwork and offense, for example, are more important than the others. To find a weighted score, you might double the numbers for teamwork and offense and add them to the values in the other categories. For example, Daugherty's weighted total would be 41 + 37 + 2(34) + 2(38) + 41 = 263 (see table 2). Complete the column "Weighted Points" using this procedure. Now select two or more categories that you feel are more important, weight them, and complete the column "Your Weighted Points"; then rank the players.

Table 2

Player	Total Points	Ranking	Weighted Points	Ranking	Your Weighted Points	Ranking
Daugherty			263			
Ewing						
Gilmore						
Jabbar						
Laimbeer						
Malone						
Olajuwon						
Parrish						
Rollins						
Sikma						

The editors wish to thank Gail Burrill, Whitnall High School, 5000 South 116 Street, Greenfield, WI 53228, for writing this issue of *NCTM Student Math Notes*.

Comparing These Methods

- How do the three methods compare? _____
- Which method did you like the best and why did you like it? _____
- Can you think of another way to rank the players? _____
- Who is your choice for the number-one center in the NBA? _____ Why? _____
- Who is the second best? _____ Why? _____
- If you were Malone's agent, which statistics would you use to stress that he was more valuable than Ewing? _____ Why? _____

Using Box-and-Whisker Plots

General managers might prefer to pay higher salaries to players who consistently have high ratings. A box-and-whisker plot of the ratings can help show consistency. To make a box plot, arrange the values for an individual player in order of size. Find the median, or middle, number and plot it on a number line. For example, Patrick Ewing's ratings are 47, 45, 39, 36, and 34, with a median of 39. Plot the lower quartile for the numbers, which is halfway between 36 and 34, or 35. Plot the upper quartile, which is 46, the average of 47 and 45. Draw a box from the lower-quartile value to the upper-quartile value (from 35 to 46). Draw a line, or whisker, from the box to the smallest rating, 34, and from the box to the largest value, 47. The line through the middle of the boxes is the middle, or median, rating. Complete the box plots below showing the range and the distribution of the five ratings given in table 1 for each player.

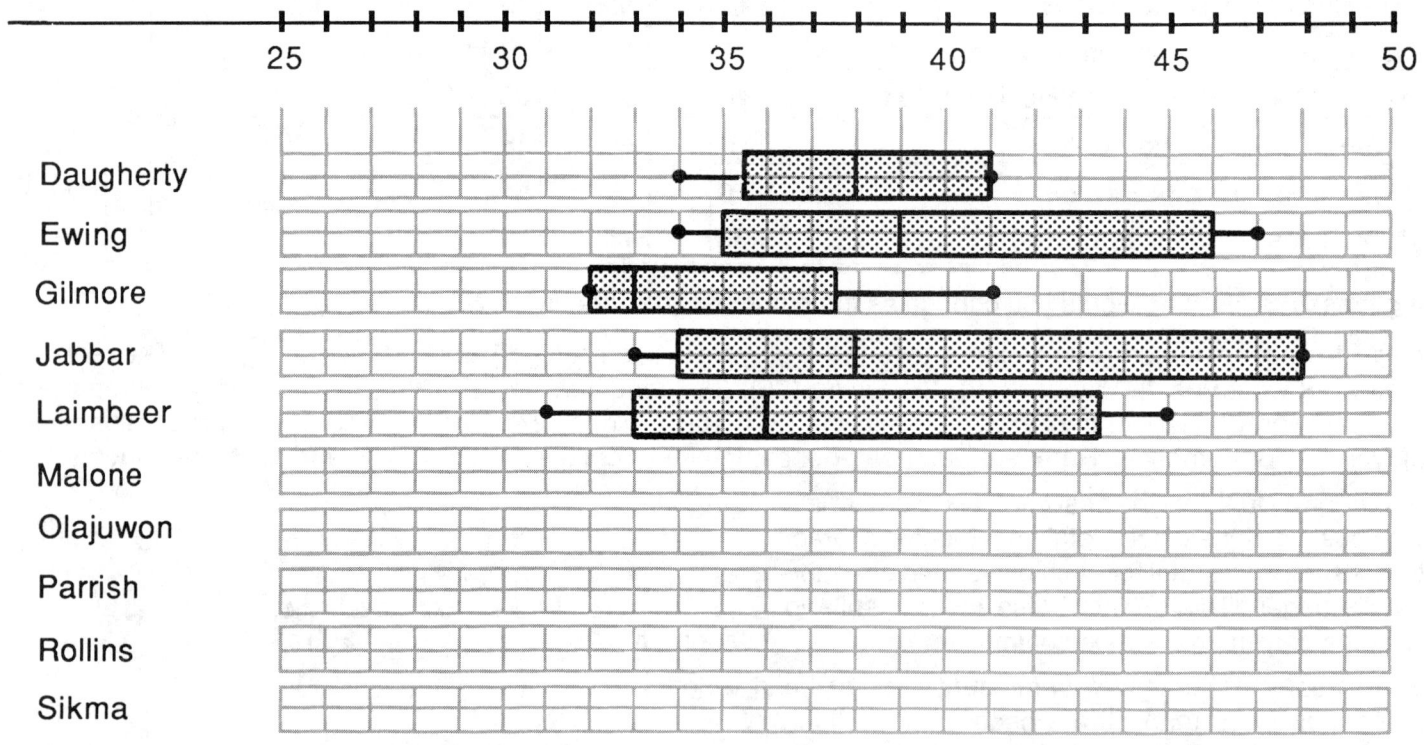

1. What is the median rating for Jack Sikma? _____
2. Which player has the lowest rating? _____
3. What players have more than half of their ratings above 37? _____
4. Which player has the highest median? _____
5. Which player is the most consistent in all his ratings? _____
6. Which player has the greatest variation in the ratings? _____
7. Which players have one rating that is much higher than their other ratings? _____
8. Sometimes the lowest score is discarded. Which player has four scores higher than any other player's four highest scores? _____

Using Scatterplots

Another way to look at the ratings is to look at combinations of the categories. Suppose you want to see who rates the highest in offense and defense only. A *scatterplot* of the offensive and defensive ratings for each player can help. For example, the point P corresponds to the ordered pair (39,40), the offensive and defensive ratings, respectively, for Robert Parrish.

Notice that the line $y = x$ has been drawn in on the scatterplot:

9. Where is the point for a player who has an equal offensive and defensive rating? _____.

10. The highest possible rating for both offense and defense is represented by the point (50, 50), since 50 is the _____ maximum possible rating in any category. The point for the player who is best in both categories should be closest to the line $y = x$ in the upper-right corner. Which center has the highest rating for both offense and defense? _____

11. Which players rate higher in offense than defense? _____
    _____
    Where are the points representing those players? _____

12. What do the players above the line have in common? _____

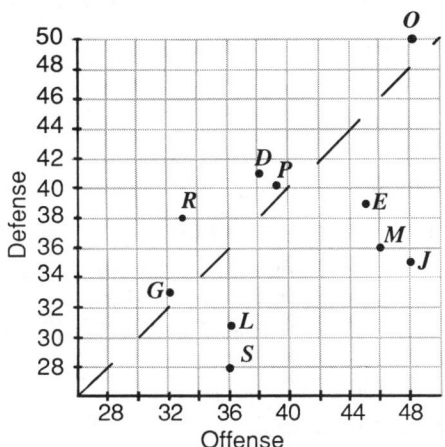

Complete the scatterplot for teamwork and offense.

13. Are the centers stronger in offense or in teamwork? _____ Why?
14. Which player is rated the highest in both offense and teamwork? _____
15. Sometimes statisticians make scatterplots of all possible pairs of categories and call this a matrix of the scatterplots. They look for any patterns they can find and to see which players rate the highest most often. How many different scatterplots would you have to make for a complete matrix of the NBA centers in which all pairs of categories are compared? _____

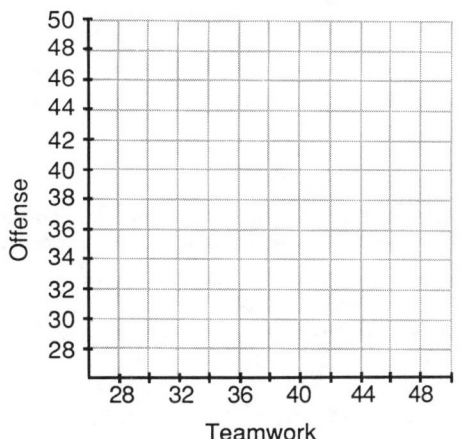

Using the Distance Formula

The distance formula can be used to find an algebraic solution for the player who has the highest rating in both offense and defense. The distance between two points (x_1, y_1) and $(x_2, y_2) =$
$$\sqrt{(x_2 - x_1)^2 + (y_2 - y_1)^2}.$$

To find the distance from $P(39, 40)$ for Robert Parrish to the point $(50, 50)$, the highest possible rating, we have

$$\sqrt{(50 - 39)^2 + (50 - 40)^2} = \sqrt{11^2 + 10^2}$$
$$= \sqrt{121 + 100}$$
$$= \sqrt{221}.$$

Robert Parrish is $\sqrt{221}$ units from the highest possible rating in both categories. You could simplify or use a calculator to approximate $\sqrt{221}$, but the numbers are easily compared in this form.

Find similar distances for each of the other players. Complete the chart for the comparison of both (offense, defense) and (teamwork, offense). Do the algebraic results using distance agree with your answers using the scatterplots?

Which method do you prefer? _____

Player	(Offense, Defense) Distance	(Teamwork, Offense) Distance
Daugherty		
Ewing		
Gilmore		
Jabbar		
Laimbeer		
Malone		
Olajuwon		
Parrish	$\sqrt{221}$	
Rollins		
Sikma		

Did You Know That . . .

- the formula for the number of possible combinations of n categories taken r at a time is $n!/(r!(n-r)!)$?
- Pat Ewing is currently the highest paid active NBA player, with a salary of $2,750,000 a year?
- the *Sporting News* 1987 Basketball Special rated Olajuwon as the number-one center, followed by Malone, Laimbeer, Ewing, and Sikma, in that order? The results were calculated by using the TENDEX formula: points plus rebounds plus assists plus blocked shots plus steals minus turnovers minus missed field-goal and free-throw attempts, with the total divided by minutes played and that quotient divided by the game pace.
- Statistics or data analysis is required for a college degree in many fields, including sociology, psychology, history, occupational therapy, accounting, economics, industrial engineering, and geography?
- the NBA center with the best 1986 free-throw shooting record is Jack Sikma of the Milwaukee Bucks, who made 87 percent of his free throws?
- the tallest NBA player is Manute Bol, who is 7 feet, 6 inches tall, and the shortest is Tyrone Bogues at 5 feet, 3 inches, both on the Washington Bullets?

Can You . . .

- find the number of different combinations if you compare three, four, or five of the categories from table 1 at a time?
- think of any other situations that could be analyzed in the same way as the NBA centers?
- rate the centers by finding the median for each category, assigning a plus 1 for a score above the median and a minus 1 for a score below the median, and then finding the total for each player?
- set up a similar rating scale for movies?

NCTM STUDENT MATH NOTES is published as part of the NEWS BULLETIN by the National Council of Teachers of Mathematics, 1906 Association Drive, Reston, VA 22091. The five issues a year appear in September, November, January, March, and May. Pages may be reproduced for classroom use without permission.

Editor: Lee E. Yunker, West Chicago Community High School, West Chicago, IL 60185
Editorial Panel: Daniel T. Dolan, Office of Public Instruction, Helena, MT 59620
Elizabeth K. Stage, Lawrence Hall of Science, University of California, Berkeley, CA 94720
John G. Van Beynen, Northern Michigan University, Marquette, MI 49855
Editorial Coordinator: Joan Armistead
Production Assistants: Ann M. Butterfield, Karen K. Aiken

Printed in U.S.A.

Teacher Notes

Page 57. In table 1, the experts' individual ratings range from a high of 10 to a low of 1, and more than one player can receive the same rating within the same category. Only one player in one category, Olajuwon in defense, received a 10 from each of the five experts for a total of 50 points. In contrast, the rankings for the ten ratings in each category range in reverse order from a high of 1 to a low of 10, and except for ties, no two players can receive the same ranking in the same category. Large differences in rating totals can show up as small differences in ranking, and vice versa.

In table 2, the same inverse relationship holds between weighted points and the corresponding rankings.

Page 58. The previous page required good number sense. The comparison questions here focus on the importance of developing good data sense. The questions show how the same set of data can be used in different ways to produce different results.

The box-and-whisker plots give a visual focus to the numerical ratings of the experts and illustrate how statistics can be used to develop spatial sense.

Page 59. Scatterplots offer another visual method of comparison, this time between pairs of categories. They can be drawn easily on graphing calculators. The distance formula requires a higher level of competence in algebra and offers an additional opportunity for calculator use.

Page 60. Students interested in sports may wish to do some corresponding data analysis with ratings of current basketball centers or other players and supply updated facts and figures for those given on the top of the page.

Page 62. The general distance formula $d = \sqrt{(x_2 - x_1)^2 + (y_2 - y_1)^2 + (z_2 - z_1)^2}$ is applicable in a three-dimensional coordinate system such as the one used on page 62 where (x_2, y_2, z_2) is $(1.000, 1.000, 1.000)$.

Answers

Page 57.

Player	Total of Rankings	Order	Total Points	Ranking	Weighted Points	Ranking
Daugherty	23	4	191	5	263	5
Ewing	20	2	201	3	282	3
Gilmore	39	10	172	10	236	10
Jabbar	21	3	202	2	298	1
Laimbeer	26	6	189	6	260	6
Malone	24	5	195	4	268	4
Olajuwon	17	1	215	1	296	2
Parrish	28	7	178	7	252	7
Rollins	37	9	174	8	238	9
Sikma	32	8	174	8	244	8

Olajuwon is the number-one center according to both the order of the total rankings and the ranking of total points. However, Jabbar is highest using the suggested weighted rankings that place double weight on teamwork and offense.

Page 58.
1. 34
2. Rollins
3. Daugherty, Ewing, Olajuwon, Jabbar
4. Olajuwon
5. Daugherty
6. Malone
7. Gilmore and Rollins
8. Daugherty

Page 59.
9. On the line $y = x$
10. Olajuwon
11. Laimbeer, Sikma, Ewing, Malone, Jabbar. Below the line $y = x$
12. They rate higher in defense than offense.
13. Offense, all points below the line
14. Jabbar
15. 10

Page 60.

Player	(Offense, Defense) Distance	(Teamwork, Offense) Distance
Daugherty	$\sqrt{225}$	$\sqrt{400}$
Ewing	$\sqrt{146}$	$\sqrt{221}$
Gilmore	$\sqrt{613}$	$\sqrt{648}$
Jabbar	$\sqrt{229}$	$\sqrt{8}$
Laimbeer	$\sqrt{557}$	$\sqrt{421}$
Malone	$\sqrt{212}$	$\sqrt{545}$
Olajuwon	$\sqrt{4}$	$\sqrt{293}$
Parrish	$\sqrt{221}$	$\sqrt{346}$
Rollins	$\sqrt{433}$	$\sqrt{650}$
Sikma	$\sqrt{680}$	$\sqrt{452}$

Page 62 (Extension). By the established criteria, the best player is the one whose distance to the point (1.000, 1.000, 1.000) is the least. They are ranked accordingly:

Rank	Player	Distance
4	Michael Jordan	0.890
2	Clyde Drexler	0.873
1	John Stockton	0.803
3	Tim Hardaway	0.885
5	Kevin Johnson	0.960

Extension—GRAPHING AND COMPARING THREE VARIABLES

Basketball statistics are recorded here as percents, in three categories. The categories are scores versus attempts for two-point field goals, for three-point shots, and for free throws. Here are the 1991–92 season figures for five of the leading guards.

		PERCENTS		
Player	Team	Field goals	Three-pointers	Free throws
Michael Jordan	Chicago	0.519	0.270	0.832
Clyde Drexler	Portland	0.470	0.337	0.794
John Stockton	Utah	0.482	0.407	0.842
Tim Hardaway	Golden State	0.461	0.338	0.766
Kevin Johnson	Phoenix	0.479	0.217	0.807

The three axes of this graph correspond to the three categories. Points are plotted on this three-dimensional graph using the percents as decimals between 0.000 and 1.000.

1. The point for Michael Jordan has already been plotted. Use the same method to plot the points for the four other players.

2. Find the distance from each plotted point to (1.000, 1.000, 1.000) using the distance formula

$$d = \sqrt{(1.000 - x)^2 + (1.000 - y)^2 + (1.000 - z)^2}.$$

Use these distances to rank the players, the best player being the one closest to the perfect value in each category, represented by the point (1.000, 1.000, 1.000). This ranking assumes, of course, that field goals, three-pointers, and free throws are equally important. Consider how the percents might be weighted according to their point values of 2, 3, and 1, respectively.

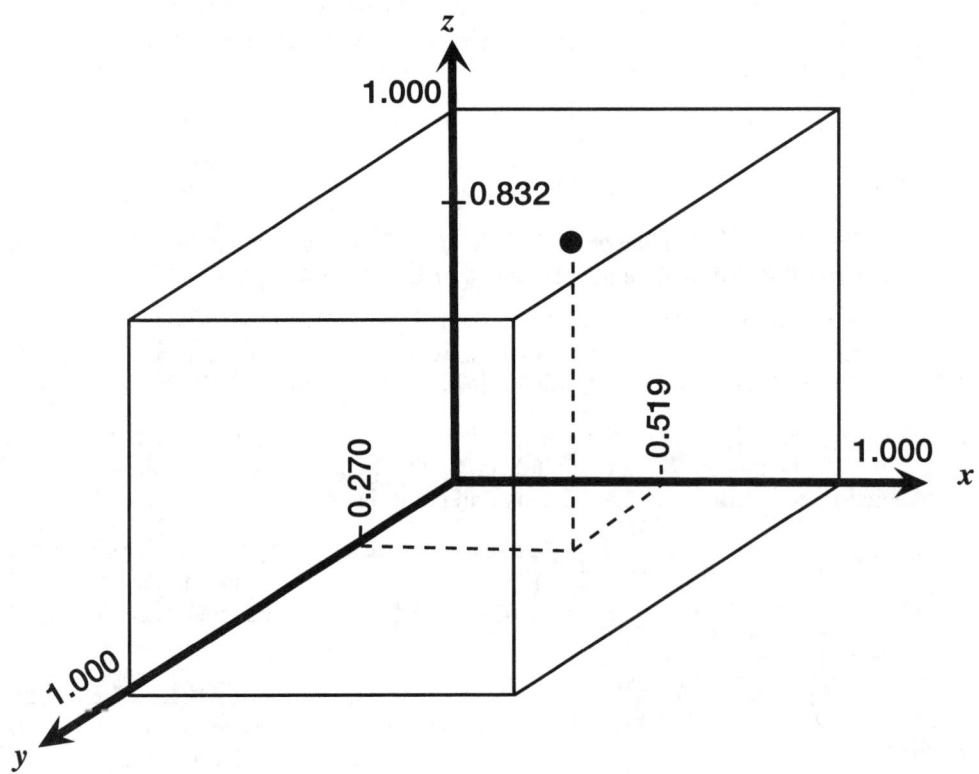

Directed Graphs

At U-Rah High School the intercom lines are connected according to the following graph:

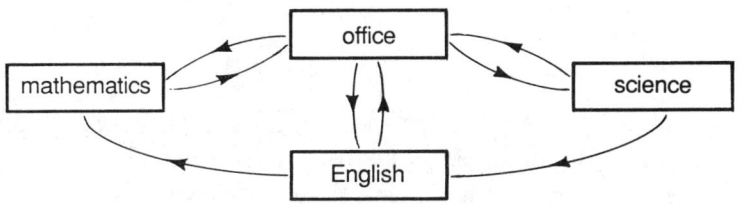

This graph is called a directed graph or *digraph,* with the arrows indicating the direction of the flow. Another method to show the same information is to use ordered pairs, where m = mathematics, e = English, o = office, and s = science.

$$(m,o), (o,m), (o,e), (o,s), (s,o), (s,e), (e,o), (e,m)$$

A third method used to show this information or to record the ordered pairs is in an array called a matrix.

	m	s	o	e
m	0	0	1	0
s	0	0	1	1
o	1	1	0	1
e	1	0	1	0

Each entry in the array corresponds to the number of single arrows connecting consecutive vertices. A 1 appears in the (e,m) cell because an arrow connects English and mathematics.

1. What is the difference between (o,s) and (s,o)? _____
2. Is it possible for the mathematics department to place a direct call to the English department? _____
3. How could the mathematics department call the English department? _____
4. What does the 1 corresponding to the pair (e,o) in the matrix mean? _____

In the diagrams below, the symbol A→B means that team A defeats team B.

5. List the ordered pairs that describe the digraph.
6. Write the matrix for the digraph.

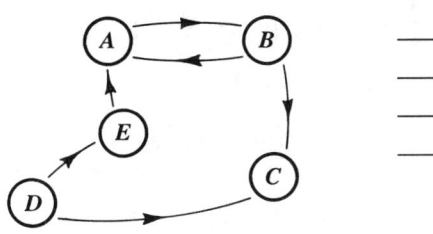
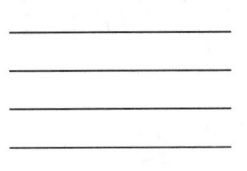
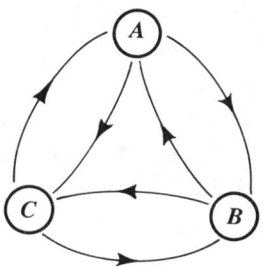

The editors wish to thank Jack and Gail Burrill, Whitnall High School, 5000 South 116 Street, Greenfield, WI 53228, for writing this issue of *NCTM Student Math Notes*.

7. From information obtained in the digraph, use the ordered pairs in number 5 to create a matrix. For example, in number 5 we see the ordered pair (E,A). From the digraph we know that E defeated A. In the matrix in the row marked E and under column A, we assign a 1. If E had not defeated A, we would have assigned a 0. Complete the matrix at the right.

	A	B	C	D	E
A	0	1			
B			1		0
C					
D					
E	1				

8. Use the matrix in number 7 to answer the following questions:
   a) What is the win-loss record of team A? _____
   b) If A is to play D, who do you predict to win? _____
   c) What is the total number of wins in the conference? _____
   d) Which team has the best record? _____
   e) Which team has the worst record? _____

At the right is a digraph concerning the diplomatic relations among certain governments.

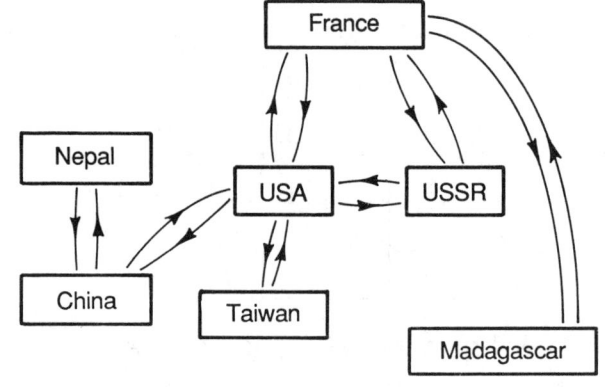

9. How would the USSR communicate with Madagascar? _____
10. How would Taiwan communicate with China? _____
11. Which countries are most isolated? _____
    Which country has the most diplomatic relations? _____
12. If every country communicated with every other country, how many ordered pairs would result? _____

Extending Your Discovery

Digraphs can be used to study sets of ordered pairs concerning specific information, such as telephone networks or tournament schedules. Some digraphs have special mathematical properties. In fact, these properties can be defined using digraphs. Let A be a set.

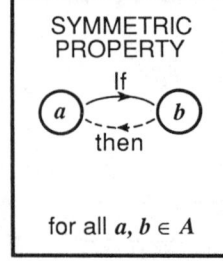

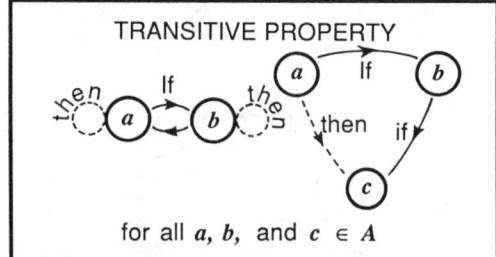

For example, the relation "is a multiple of" on the set $\{1, 2, 3, 4, 6\}$ results in the digraph at the right. This relation has both the reflexive and transitive properties. It does not have the symmetric property because 6 is a multiple of 3 but 3 is not a multiple of 6.

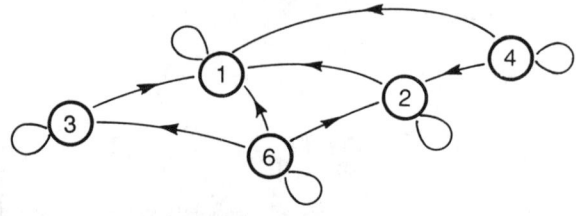

$a \rightarrow b$ means a is a multiple of b

13. Which properties listed on page 64 do the following relations have on the given sets? (The corresponding digraphs are pictured.)

   a) "Is taller than" on the set of children listed below:

   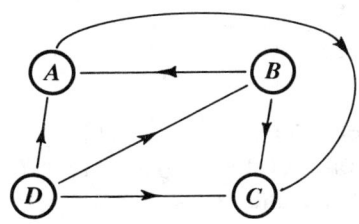

Name	Height
Anna	85 cm
Barta	90 cm
Cara	72 cm
Dona	110 cm

   b) "Is in the same grade as" on the set of children listed:

   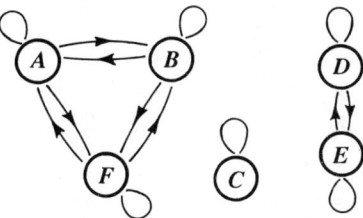

Name	Grade
Anna	1
Bai	1
Casey	3
David	2
Evan	2
Francis	1

   c) "Is a factor of" on the set {1, 2, 4, and 8}

   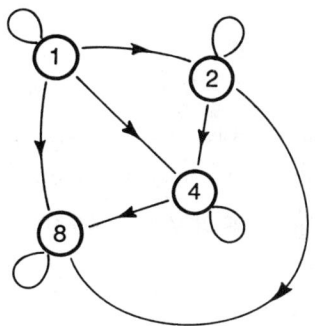

14. Decide whether each relation at the right has reflexive, symmetric, or transitive properties.

   a)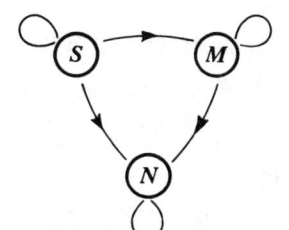

   b)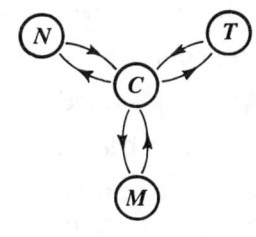

15. In the digraph shown, involving twelve people, part of the arrows are drawn on the relation "is the sister of." Add any other arrows that can be deduced.

   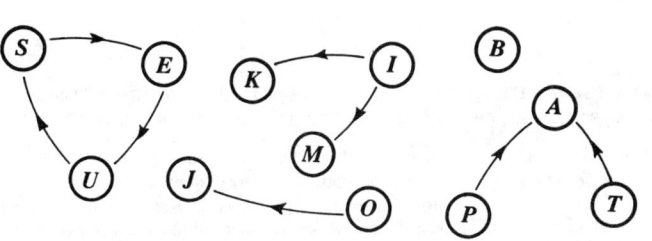

16. In the digraph pictured below, $a \rightarrow b$ means a is the mother of b.

    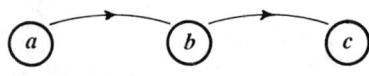

    a) How is a related to c? _____
    b) How is b related to a? _____
    c) How is c related to b? _____
    d) How is c related to a? _____

17. Make a digraph for each of these sets of ordered pairs and decide whether each relation has reflexive, symmetric, or transitive properties.

    a) {(1,1), (1,2), (1,3), (2,1), (2,3), (2,2), (3,3), (3,1), (3,2)}
    b) {(1,2), (2,3), (1,3), (3,1), (2,2), (3,2)}
    c) {(1,1), (1,3), (2,3), (3,2), (3,1), (2,4), (3,4), (4,4)}

18. Decide whether each relation has reflexive, symmetric, or transitive properties.

    a)
	1	2	3
1	0	1	1
2	1	1	0
3	0	1	1

    b)
	1	2	3
1	1	1	1
2	1	1	1
3	1	1	1

Can You . . .

- find the number of intercom lines you would need to connect each of five people to each other?
- find a relation that has the reflexive and symmetric properties but not the transitive properties?
- find a relation that has the symmetric property but not the transitive property?

Did You Know That . . .

- digraphs are derived from the work done by Leonhard Euler and the Königsberg bridge problem?
- digraphs can be used to find the critical path showing the length and sequence of the activities in a project?
- digraphs were used to help develop the System Flow Plan for the engine of the Polaris submarine?
- digraphs can be used to trace the passing of DDT through the ecological food chain?
- the square of a matrix for a digraph will give the number of two-edged paths that connect any two vertices; the cube will give the number of three-edged paths that connect any two vertices?

NCTM STUDENT MATH NOTES is published as part of the NEWS BULLETIN by the National Council of Teachers of Mathematics, 1906 Association Drive, Reston, VA 22091. The five issues a year appear in September, November, January, March, and May. Pages may be reproduced for classroom use without permission.

Editor: Daniel T. Dolan, Office of Public Instruction, Helena, MT 59620
Editorial Panel: Johnny W. Lott, University of Montana, Missoula, MT 59812
Judy Olson, Western Illinois University, Macomb, IL 61455
John G. Van Beynen, Northern Michigan University, Marquette, MI 49855
Editorial Coordinator: Joan Armistead
Production Assistant: Ann M. Butterfield

Printed in U.S.A.

Teacher Notes

Page 63. The directed graph, ordered pairs, and matrix all show the same interconnections. Direction is visually apparent in the digraph but implied in the ordered pairs and matrix, being read from first to second and from row to column, respectively.

Page 64. You may wish to have students construct a digraph, ordered pairs, and matrix for part of a playoff series in baseball, football, or basketball. Have students discuss how the digraph for the relation "is a multiple of" would change if the relation were "is a factor of" and "is a prime factor of."

Page 66. Questions 17 and 18 require the recognition of the reflexive, symmetric, and transitive properties through ordered pairs and matrices. Encourage students to express the process they use verbally.

Answers

Page 63.
1. (*o,s*) indicates that the office can call the science department directly; (*s,o*) indicates that the science department can call the office directly.
2. No
3. One possibility is (*m,o*) and (*o,e*).
4. An arrow connects the English department to the office.
5. (A,B), (B,A), (B,C), (D,C), (D,E), (E,A)

6.
	A	B	C
A	0	1	1
B	1	0	1
C	1	1	0

Page 64.
7.
	A	B	C	D	E
A	0	1	0	0	0
B	1	0	1	0	0
C	0	0	0	0	0
D	0	0	1	0	1
E	1	0	0	0	0

8. (a) Won 1, lost 2; (b) Because A has won one game while D has won two, you might expect D to win. Also D defeated E and E defeated A, so that you might expect D to win; (c) 6; (d) B and D have each won two games. (e) C has won no games.
9. One way is through France to Madagascar
10. One way is through USA to China
11. Nepal, Madagascar; USA
12. 42

Page 65.
13. a) The transitive property
    b) The reflexive, symmetric, and transitive properties
    c) The reflexive and transitive properties
14. a) Reflexive and transitive
    b) Symmetric

15.

Page 66.
16. a) *a* is the grandmother of *c*.
    b) *b* is the daughter of *a*.
    c) *c* is the child of *b*.
    d) *c* is the grandchild of *a*.
17. a) The relation is reflexive, symmetric, and transitive.   b) None   c) None

18. (a) None; (b) all

Page 68 (Extension).

1.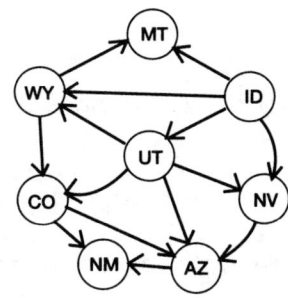

2.
	AZ	CO	ID	MT	NM	NV	UT	WY
AZ	0	0	0	0	1	0	0	0
CO	1	0	0	0	1	0	0	0
ID	0	0	0	1	0	1	1	0
MT	0	0	0	0	0	0	0	0
NM	0	0	0	0	0	0	0	0
NV	1	0	0	0	0	0	0	0
UT	0	0	0	0	1	1	0	1
WY	0	1	0	1	0	0	0	0

3. Five states: (ID, UT), (UT, NV), (NV, AZ), (AZ, NM)

Extension—TRAVELING WITH DIRECTED GRAPHS

The areas of the mountain and western states of Arizona, Colorado, Idaho, Montana, Nevada, New Mexico, Utah, and Wyoming are given in square miles in the chart below:

AZ	113 909
CO	104 247
ID	83 557
MT	147 138
NM	121 666
NV	110 540
UT	84 916
WY	97 914

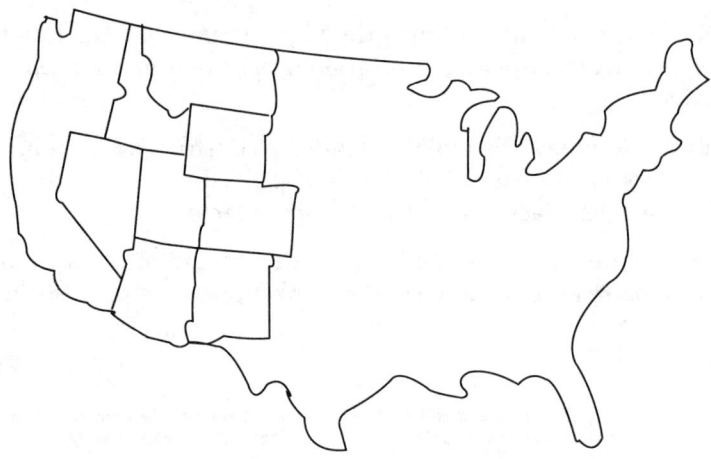

1. Insert the necessary arrows to complete this directed graph, using "travel to a larger border state" as the relation among this set of states. Assume two states are border states if they share at least one border point in common.

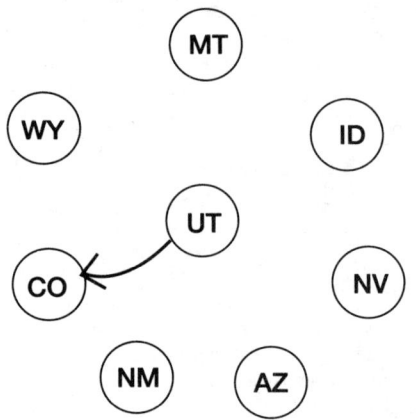

2. Use the same information to complete the corresponding matrix. Use 0 if no path exists and 1 if a path between the states does exist.

	AZ	CO	ID	MT	NM	NV	UT	WY
AZ								
CO								
ID								
MT								
NM								
NV								
UT								
WY								

3. What is the greatest number of states that can be traveled in sequence under this relation? Give the trip as a sequence of ordered pairs.

4. Try collecting area data from another part of the country, drawing a similar directed graph and completing the corresponding matrix.

NATIONAL COUNCIL OF TEACHERS OF MATHEMATICS   NOVEMBER 1988

Investigating Perimeter and Area

DeCora and DeCore, interior decorators, are constructing mosaic designs using colored ceramic square tiles. Each square is the same size, and the length of a side is one unit. With six squares, they created the following designs:

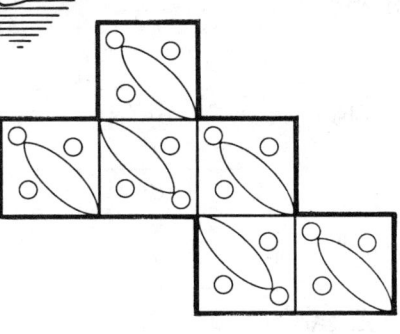

Design 1

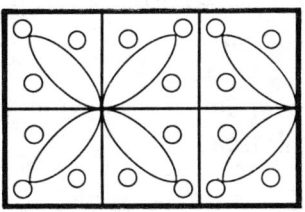

Design 2

It is interesting to note that even though the areas are the same, the perimeters are different. Find the perimeter of each figure to see how they differ. To determine the price of a mosaic, DeCora and DeCore charge $8 for each tile and $5 per unit of length for framing (1 unit of length = the length of one side of the tile).

- Find the total cost of the tile and framing for each mosaic shown above.

    Design 1 = _____   Design 2 = _____

DeCora and DeCore often have to explain to customers that designs with the same area may have different total costs, so they need a good understanding of the relationship between the perimeter and area of polygonal figures. The following activities offer an opportunity to explore the relationship between perimeter and area using arrangements of squares. In each arrangement, neighboring squares must share a common side.

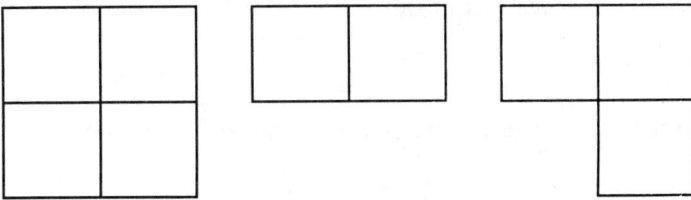   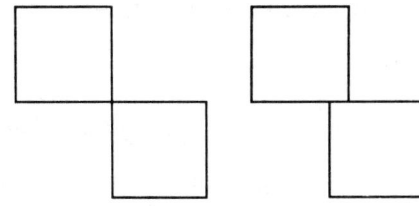

(These arrangements can be used.)   (These cannot be used.)

Investigating Perimeter

You will need grid paper and fifty to one hundred square tiles or squares cut from construction paper. Let the length of the side of a square be one unit. Note that the area of each arrangement is equal to the number of squares. Work with a group of classmates to investigate the perimeter of certain arrangements found by putting squares together.

The editors wish to thank Judy Mumme, mathematics department of the University of California at Santa Barbara, for writing this issue of *NCTM Student Math Notes*.

Investigating Perimeter—Continued

- Verify the results in table 1 for one, two, and three squares.

- Explore different arrangements of squares to determine the perimeter for each.

- Explore the range of perimeters possible for each number of squares to find the arrangements that produce the maximum and minimum perimeters.

- Use grid paper to sketch the arrangements for the maximum and minimum perimeters.

- Record the maximum and minimum perimeters for the area arrangements in table 1.

- Look for patterns in the table and on the grid paper to see if you can predict the maximum and minimum perimeter of an arrangement. You should extend table 1.

Table 1

Area (No. of Sq.)	Maximum Perimeter	Minimum Perimeter
1	4	4
2	6	6
3	8	8
4	10	
5		10
6		
7		
8		
9		
10		
11		
12		
13		
14		
15		
16		
17		
18		
19		
20		

1. Can you find an arrangement for which the perimeter is an odd number? If so, draw the arrangement. If not, explain. _____

    _____

    _____

2. Predict the *maximum* perimeter of arrangements of the following numbers of squares:

    40 _____   75 _____   100 _____

3. Make a conjecture for finding the *maximum* perimeter of arrangements of any number of squares.

    _____

    _____

4. Now, try to predict the *minimum* perimeter of arrangements of the following numbers of squares:

    36 _____   48 _____   55 _____

5. Make a conjecture for finding the *minimum* perimeter of arrangements of any number of squares.

    _____

    _____

Investigating Areas

Using the same rule for arranging squares, explore the minimum and maximum *area* of an arrangement of squares with a *given perimeter*.

- The arrangement shown in the grid has a perimeter of 10. Outline five other arrangements that have a perimeter of 10. Indicate the area of each arrangement on the grid.

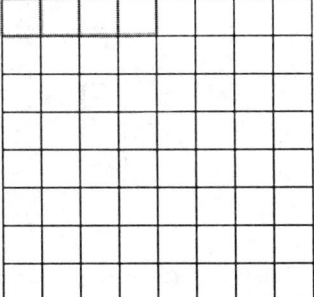

- The arrangement shown on this grid has a perimeter of 12. Outline as many other arrangements as you can that have a perimeter of 12. Indicate the area of each.

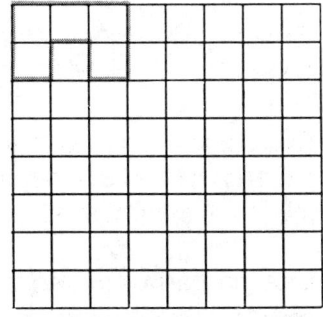

- Continue to explore different arrangements of squares that produce a given perimeter and determine the area of each arrangement in order to find arrangements that produce the *minimum* and *maximum* areas. Record these values in table 2.

Table 2

Perimeter	Minimum Area	Maximum Area
4	1	1
6	2	2
8	3	4
10	4	
12		9
14		
16		
18		
20		
22		
24		
26		

6. Predict the *minimum* area for the following perimeters:

   30 _____   36 _____   44 _____

7. Make a conjecture for finding the *minimum* area.

   _____
   _____
   _____

8. Predict the *maximum* area for the following perimeters:

   30 _____   36 _____   42 _____

9. Make a conjecture for finding the *maximum* area.

   _____
   _____
   _____

Teaching with *Student Math Notes*: Volume 2

Did You Know That...

- using the same rule as before, only one way exists to arrange one square, one way to arrange two squares, two ways to arrange three squares, and five ways to arrange four squares?

- you can add squares to an arrangement to increase the area and decrease the perimeter?
- this investigation can be extended to three dimensions using cubes to explore the relationship between surface area and volume?
- design and packaging engineers also have to consider the relationship between perimeter and area in maximizing profit while minimizing cost. Similar concerns face civil engineers and housing developers when they lay out streets and lot designs.

Can You...

- find the number of ways to arrange five squares? Six squares?
- find various arrangements of squares whereby adding a square will decrease the perimeter?
- find a rule for maximum and minimum perimeters for a given number of equilateral triangles?
- find a shape that will produce the maximum area for a given perimeter?
- find the arrangements of five squares that will fold into an open-top box?
- find an arrangement of cubes whereby adding a cube to an arrangement increases the volume and decreases the surface area?

NCTM STUDENT MATH NOTES is published as part of the **NEWS BULLETIN** by the National Council of Teachers of Mathematics, 1906 Association Drive, Reston, VA 22091. The five issues a year appear in September, November, January, March, and May. Pages may be reproduced for classroom use without permission.

Editor:	**Dan Dolan,** Office of Public Instruction, Helena, MT 59620
Editorial Panel:	**Johnny W. Lott,** University of Montana, Missoula, MT 59812
	Judy Olson, Western Illinois University, Macomb, IL 61455
	John G. Van Beynen, Northern Michigan University, Marquette, MI 49855
Editorial Coordinator:	**Joan Armistead**
Production Assistants:	**Ann M. Butterfield, Sheila C. Gorg**

Printed in U.S.A.

Teacher Notes

Page 69. The hands-on aspects of this activity are very important. Be sure students understand the rules for arranging the squares. Common sides must be joined in their entirety and no squares can overlap.

Page 70. The number patterns developed here for perimeter may prove challenging for some students. Encourage cooperative efforts. Choose simpler values for the specific numerical cases, and offer helpful hints on the generalizations if necessary.

Page 71. Here again, adapt the questions on areas to fit the ability levels of your students.

Page 74. In this activity, count only cuts that produce different shapes as different. Do not count different orientations of cuts that produce the same shapes as others already found.

Answers

Page 69. $118, $98

Page 70. **Table 1**

Max: 4 6 8 10 12 14 16 18 20 22 24 26 28 30 32 34 36 38 40 42 44 46 48 50 52 ...
Min: 4 6 8 8 10 10 12 12 12 14 14 14 16 16 16 16 18 18 18 18 20 20 20 20 20 ...

1. No, the perimeter of a square is an even number. If a square is added to an arrangement, the perimeter stays the same or increases or decreases by a multiple of 2.
2. 82, 152, 202
3. Twice the number of squares plus 2. $P = 2n + 2$
4. 24, 28, 30
5. Answers will vary with the sophistication of the student.
   Let a be the number of squares and g be the greatest integer on $\sqrt{a}$, the whole-number square root.
   If $\sqrt{a} = g$, then $p = 4\sqrt{a}$.
   If $a \leq g^2 + g$, then $p = 4g + 2$.
   If $a > g^2 + g$, then $p = 4(g + 1)$.

Page 71. **Table 2**

Max: 1 2 3 4 5 6 7 8 9 10 12 14 16 18 20 ...
Min: 1 2 4 6 9 12 16 20 25 30 36 42 49 56 64 ...

6. 14, 17, 21
7. Minimum area: Subtract 2 from the perimeter and divide the result by 2. $4 = (p - 2)/2$.
8. 56, 81, 110
9. Maximum area: Let q equal the greatest integer in $P/4$, the whole-number quotient of $P/4$.
   If P is a multiple of 4, then the maximum area is q^2.
   Otherwise the maximum area is $q(q + 1)$.

Page 74 (Extension).   1. Perimeters: 18   2. Perimeters: 38

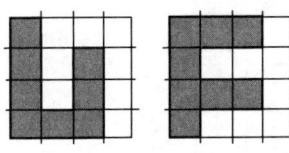
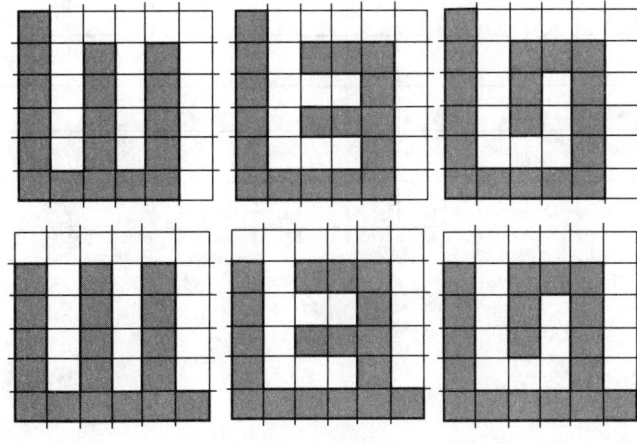

Extension—MAXIMIZING PERIMETERS

There is only one way, except for orientation, to cut a 2 × 2 square into two congruent halves by cutting on the unit grid lines. The number of square units of area in each piece is half the original square. But the number of units of perimeter in each piece is three-fourths that of the original square.

1. There are six different ways to cut a 4 × 4 square into two congruent halves by cutting on the unit grid lines. Draw the two different cuts that give the two pairs of congruent shapes with the greatest possible perimeter.

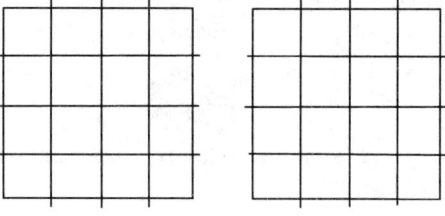

Perimeter: _____

2. There are many different ways to cut a 6 × 6 square into two congruent halves by cutting on the unit grid ines. Draw six different cuts that all give two congruent pieces with the greatest possible perimeter.

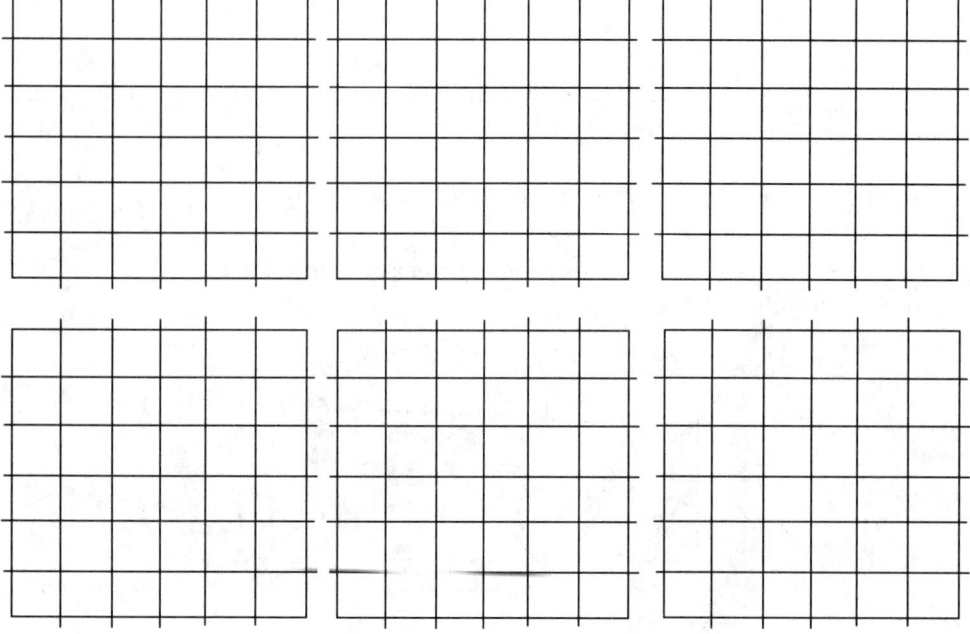

Perimeter: _____

NATIONAL COUNCIL OF TEACHERS OF MATHEMATICS — JANUARY 1989

Star Rec

It was A.D. 8128—a perfect year for star reckoning. Captain Turk and Spec, his intuitive aide, were gazing at a star in the Circle galaxy. Turk, gloating because his telescope was more powerful than Spec's, decided to stump Spec with a question. "Oh, Spec, what is the sum of the five angles in the tips of star Alpha-5X?"

Spec peered through his tiny telescope and immediately announced his answer. It was extremely close to the answer that Turk had calculated with his protractor. Spec added that the answer was "obvious" to any "subsimian with eyes."

Turk, who had spent some time measuring the angles at the tips, was astonished that the result was obvious to Spec. Let's explore their methods to see how each arrived at his result.

Turk's Method—Measuring

In his powerful telescope, Turk saw Alpha-5X (fig. 1).

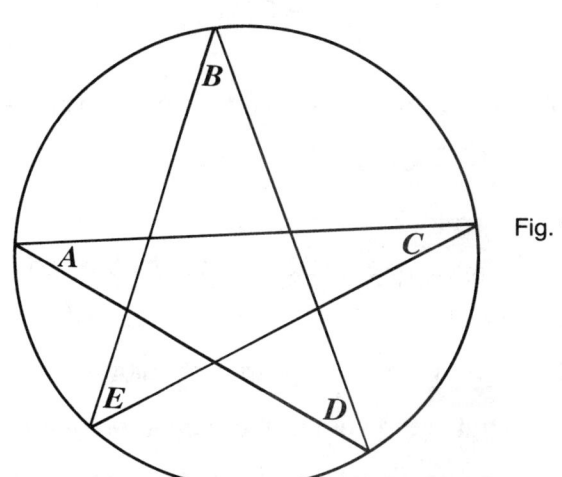

Fig. 1

1. With a protractor, measure angles A, B, C, D, and E. What is the sum of the five angle measures? _____
2. Compare your results with those of your classmates. If your answers differ, how can you resolve the differences so that you get an answer that is acceptable to the entire class? _____
3. What angle sum did you agree on? _____

Measure the lettered angles in the tips of the stars in figures 2 and 3. Write the measure of each angle by its vertex. Then compute the sums of the measures of the angles.

4. Angle sum = _____

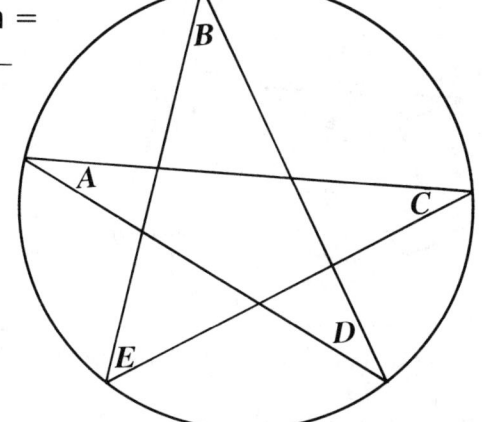

Fig. 2. **Alpha-5Y**

5. Angle sum = _____

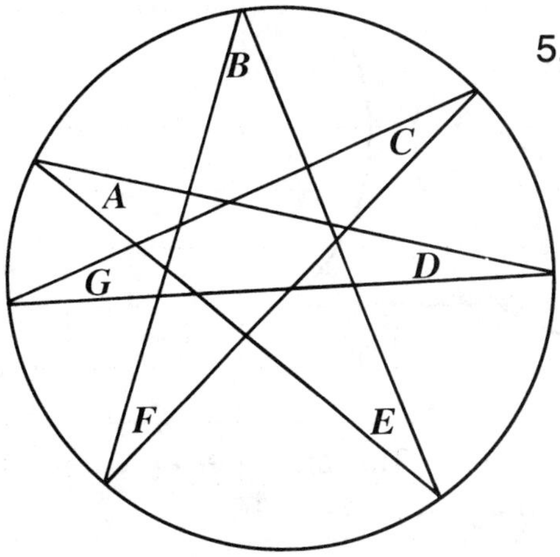

Fig. 3. **Alpha-7X**

6. What generalization can you make about the sum of the angle measures in stars in the Circle galaxy?

The editors wish to thank Maurice Burke, Department of Mathematical Sciences, Montana State University, Bozeman, MT 59715, for writing this issue of *NCTM Student Math Notes*.

Spec's Method—Zooming Out

In Spec's less powerful telescope, stars appeared very small as in figure 4. Spec visualized light rays emanating from the tips and radiating to the boundary of the telescope lens. These light rays were sides of the star's angles or sides of congruent vertical angles, as shown in figure 5. Using the zoom feature on his telescope, Spec "zoomed out" until Alpha-5X appeared as in figure 6. The star seemed to grow smaller, but the light rays still extended to the edge of the telescope lens.

Fig. 4

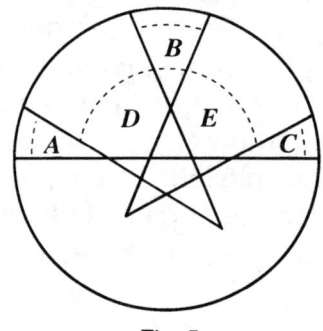
Fig. 5

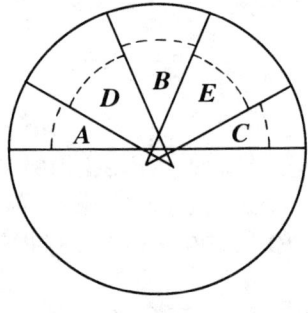
Fig. 6

7. What appears to have happened to the vertices of the angles to be measured as Spec zoomed out on the star?

8. What appears to be formed by the two sides of angles A and C nearest the bottom of the telescope lens?

9. What appears to be true about the sum of the measures of angles $A, B, C, D,$ and E? _____

10. What is the sum of the measures of the angles of the star? _____

Draw in appropriate angles in the views of Alpha-5Y and Alpha-7X shown in figures 7 and 8. Determine if Spec's method of "zooming out" results in the same angle sum for these stars.

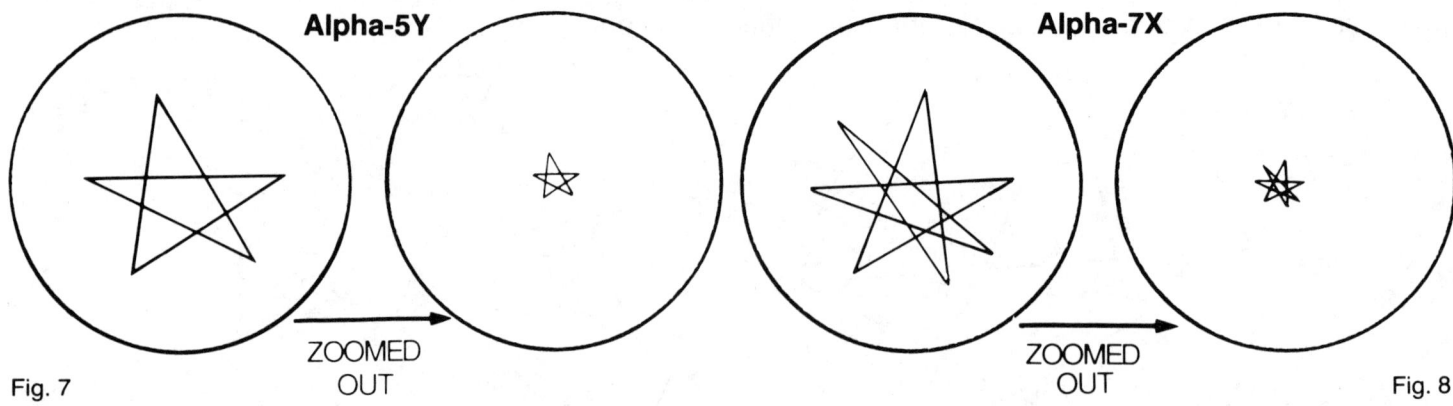
Fig. 7 — Alpha-5Y ZOOMED OUT — Alpha-7X ZOOMED OUT — Fig. 8

11. After much discussion, Turk and Spec agreed that the answer was 180 degrees but that their methodology left some questions unanswered. Why?

Extensions of Turk's and Spec's Discovery

After having discovered two interesting theorems in an ancient twenty-first-century A.D. manuscript entitled "The Lost Lore of Circles," they focused on the circular boundary of their telescopic views to prove their result.

Teaching with *Student Math Notes:* Volume 2

Excerpt from "The Lost Lore of Circles"

INSCRIBED ANGLE THEOREM: *The measure of an angle inscribed in a circle is equal to one-half the measure of its intercepted arc.*

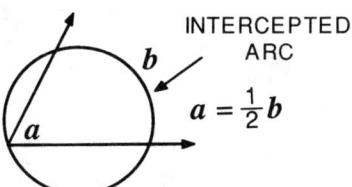

ANGLE THEOREM FOR CHORDS: *The measure of an angle formed by two chords intersecting within a circle is equal to one-half the sum of the measures of the arcs intercepted by the angle and its vertical angle.*

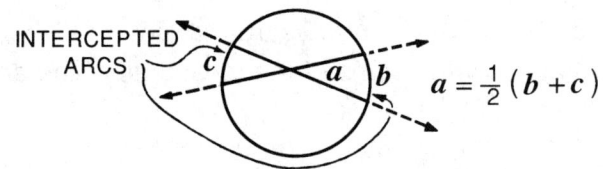

Turk's reasoning: Assume that the view boundary in figure 9 is a circle and that angles $A, B, C, D,$ and E are formed by the intersecting chords. By using the inscribed angle theorem, one can determine that the sum of the five angles is one-half the entire circle—360°/2, or 180°. Why?

Fig. 9

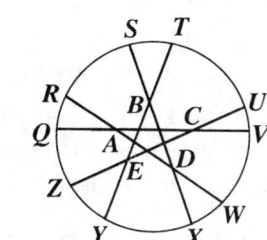

Spec's reasoning: Assume that the view boundary in figure 10 is a circle and that the angles of the star at $A, B, C, D,$ and E are formed by chords. Using the angle theorem for chords, one can show that the sum of the five angles at the tips of the stars is one-half the sum of the measures of the ten intercepted arcs, or one-half the entire circle. Thus the sum is 360°/2, or 180°. Why?

Fig. 10

Spec pointed out that Turk's proof relied on the assumption that the tips of the star all lie on a circle, whereas his did not. Spec further speculated that his proof would hold for any star as long as the tips were formed by secants that intercepted nonoverlapping arcs of the circle. You can investigate his conjecture in the following exercises.

12. Form a star by connecting the points numbered 1, 3, 5, 7, 2, 4, 6, and 1 in that order in figure 11. Using Spec's method, compute the sum of the measures of the angles at the tips of the star you formed.

    _____

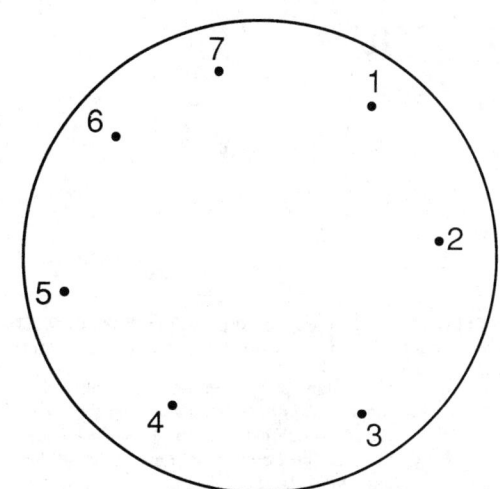

Fig. 11

13. Connect points 1, 4, 7, 3, 6, 2, 5, and 1 in that order in figure 11 using a different color. Using Spec's method, compute the sum of the measures of the angles at the tips. _____

14. Make a conjecture concerning the sum of the measures of the angles at the tips of any star. _____

Turk responded, "Not bad, Spec, but—

Can you . . .

- define 'star' so that Turk and Spec's result is always true?"
- find a result that is always true for any 'star'?"
- explain why the sums of the measures of the angles in 'stars' with an even number of points always seem to exceed 180 degrees?"
- describe how to draw a nine-pointed star such that the sum of the measures of the angles at the tips of that star will be 180 degrees?"
- write a Logo program that lets the computer 'zoom out' and 'zoom in' on a picture of a star in such a way that selected angles always extend to the edge of the screen?"

Spec replied, "Good questions, Turk. They are worthy of thought. But—

Did you know that . . .

- even though 'zooming out' does not 'prove' anything, it can help you intuitively to 'see' and discover theorems about the sums of the exterior and interior angles of polygons as well as theorems about transversals cutting parallel lines?"
- 'zooming out' is like 'dilating' the plane, and its validity as a method of argument rests on an assumption equivalent to Euclid's parallel postulate; that is, there exist polygons that are 'similar' but not congruent?"
- before 1600 B.C. the Babylonians were aware (corollary of the inscribed angle theorem) that an angle inscribed in a semicircle was a right angle? This fact is called the 'theorem of Thales' even though Thales was born over a thousand years later."

Captain Turk answered, "I think you're going to be a famous star like me one of these days." Spec dryly commented, "I reckon."

Teacher Notes

Page 75. Small errors will occur in measuring the angles with a protractor, so answers will, at best, be approximate. However, the exact sum of the degree measures of the five angles, from a theoretical point of view, must be 180.

Another informal way to establish this sum is by sweeping a pencil successively through all the five angles at the points of the star. In figure 1, place the pencil on AC pointing to the right. Pivot at A and rotate through angle A. From there, pivot at D and rotate through angle D. Continue pivoting and rotating through angles B, E, and D, in that order. At the end, the pencil will again be on segment AC. But look closely. It will be pointing the other way, to the left, having rotated in all through 180 degrees.

Page 76. In zooming out, the angles stay the same but their vertices appear to converge. Only in the limit will the entire star become a single point. If that point is on the diameter, then the intercepted arcs add to 180 degrees, but this is not a necessary requirement for the argument given.

Page 80. A three-dimensional model of a regular dodecahedron and a stellated dodecahedron would be useful as teaching aids here. However, the initial activity should be done with the drawing, where the student is required to visualize the actual solid mentally. Use the model at the end of the activity for reinforcement and check.

Answers

Page 75.
1. The measures are only approximate but should be close to the following in degrees: 33, 37, 26, 40, and 45. The sum is approximately 181 degrees.
2. Perhaps average the results
3. Approximately 180 degrees
4. Approximately 179 degrees
5. Approximately 178 degrees
6. They appear to sum to a number close to 180 degrees.

Page 76.
7. They appear to have converged to a point.
8. They appear to form a straight line.
9. They appear to form a straight angle.
10. 180 degrees
11. Their methods suggest that 180 degrees is a good answer, but they don't guarantee that the sum must be 180 degrees. A proof is needed.

Page 77.
12. 540 degrees
13. 180 degrees
14. The sum always appears to be a multiple of 180 degrees. Whenever the tips of the stars are formed by chords that intercept nonoverlapping arcs of a circle, the sum of the measures of the angles is 180 degrees.

Page 80 (Extension). 2. Six stars can be seen in the figure. The entire solid has twelve stars on its faces.

3. The isosceles triangles have one angle of 36 degrees and two of 72 degrees.
   There are twelve points on the solid, each formed by five triangular faces coming together. Thus the total number of triangular faces on the entire solid must be $12 \times 5 = 60$.
4. Eleven of the twelve points of the solid can be seen in the figure.
5. A minimum of five colors are needed.

Extension—STARS IN 3-D

This solid is called a stellated dodecahedron. It is built around a regular dodecahedron with twelve congruent pentagons as faces. The five sides of each pentagon are extended to form five-pointed stars. Formed this way, the solid appears to have congruent pentagonal pyramids emerging from its surface.

1. Imagine this solid made from interlocking stars such as the two already shaded. Look for another star and shade it with a different pattern.

2. Each star is in a different plane. How many stars can you see in the drawing? How many stars do you think are in the entire solid?

3. Each point of a star is a triangle, and all such triangles are congruent. How many such triangles are in the entire solid? What are the measures of the three angles in each triangle?

4. Look for a pyramid with a pentagon as its base. How many such pyramids or parts of pyramids can you see in the drawing? How many such pentagonal pyramids do you think are in the entire solid?

5. Assume that the five triangles forming a star are always painted in the same color and that no two stars with joining edges are painted the same color. What is the minimum number of different colors needed to paint the entire stellated dodecahedron this way?

What Is a Curve of Constant Width?

Why is a sewer cover round? A good sewer cover is round so that it won't fall through the sewer hole regardless of how it is positioned. A sewer cover is in the shape of a circle that has a constant diameter, or is of *constant width*.

Trace each figure below and determine whether it would make a good sewer cover. Place each figure between the parallel lines L_1 and L_2 on table 1. Can you rotate it such that each line L_1 and L_2 will always touch the figure at one point? If you can, the width is constant and the figure is called a *curve of constant width*.

Check each figure and indicate in table 1 whether each one is a curve of constant width.

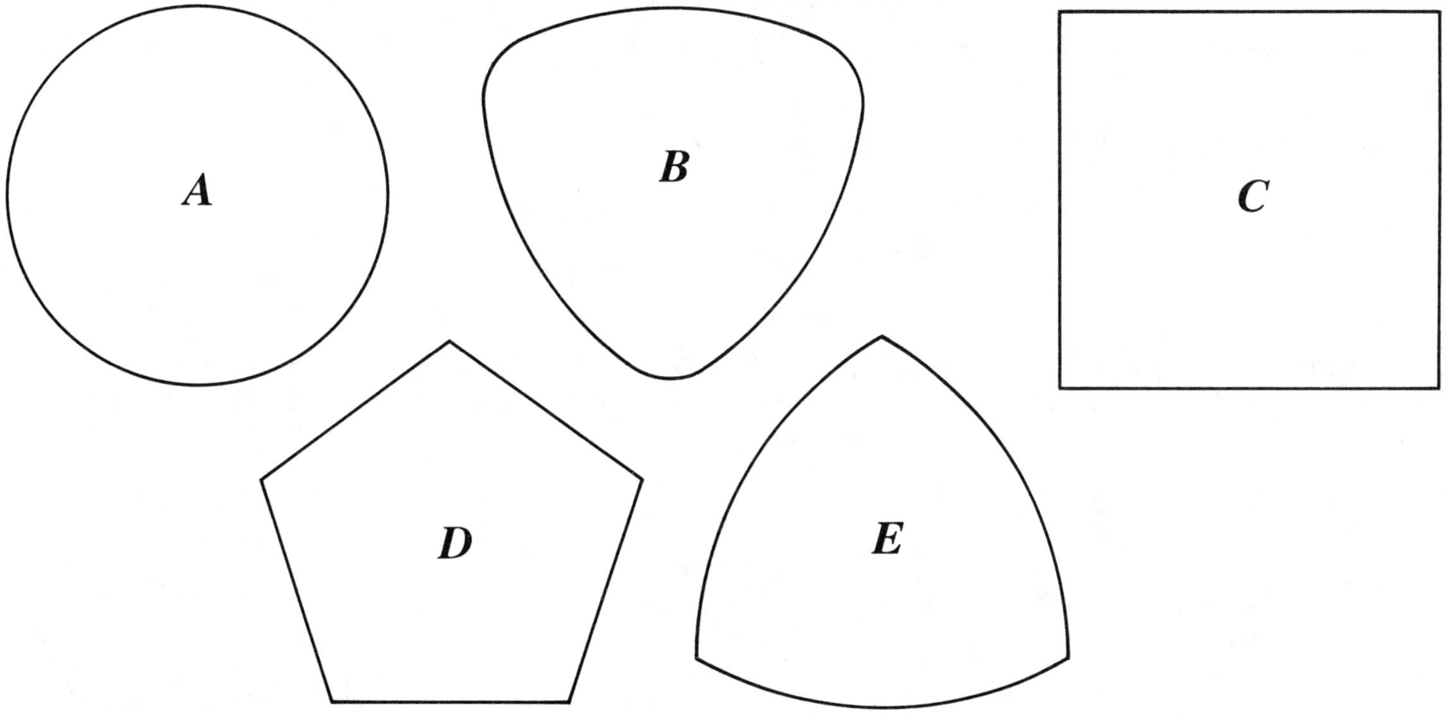

L_1

TABLE 1

Figure	A	B	C	D	E	F	G	H	I
Yes/No									

Figures F, G, H, and I can be found at the bottom of page 2.

L_2

The panel wishes to thank Millie Johnson, mathematics department of Western Washington University, Bellingham, WA 98225, for writing this issue of the *NCTM Student Math Notes*.

Constructing Curves of Constant Width

Use equilateral triangle *ABC* with sides length *h*. Open a compass to length *h*, place the point of the compass at *A*, and draw $\widehat{BC}$. Next, place the point at *B* and draw $\widehat{AC}$, then at *C* and draw $\widehat{AB}$. This figure is called a *Reuleaux* (roō-lō′) *triangle*.

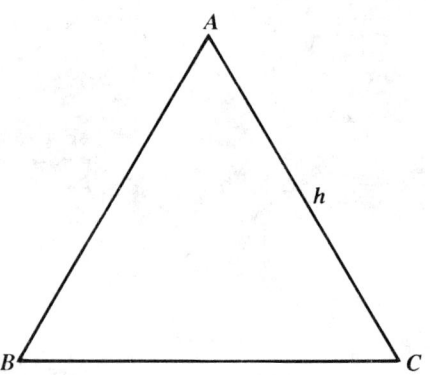

Trace the curve, cut it out, and rotate it between lines L_1 and L_2 on table 1 to verify that the Reuleaux triangle is a curve of constant width.

1. What is the distance from vertex *A* to any point on $\widehat{BC}$? _____
2. What is the distance from vertex *B* to any point on $\widehat{AC}$? _____
3. What is the distance from vertex *C* to any point on $\widehat{AB}$? _____

More Constructions

ABCDE is a regular pentagon. Let *AD* = *h*; draw $\overline{AD}$, $\overline{AC}$, $\overline{BD}$, $\overline{BE}$, and $\overline{CE}$.

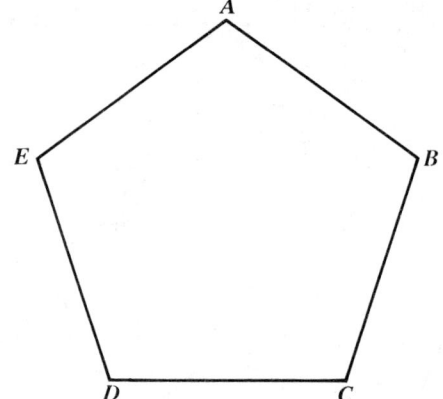

4. How are *AD*, *AC*, *BD*, *BE*, and *CE* related? _____

How can you show that this relationship is true? _____

Open a compass to length *h*. Place the point at *A* and draw $\widehat{DC}$, at *B* and draw $\widehat{DE}$, at *C* and draw $\widehat{AE}$, at *D* and draw $\widehat{AB}$, and at *E* and draw $\widehat{BC}$. This figure is called a *Reuleaux pentagon*. Trace it, cut it out, and rotate it between lines L_1 and L_2 on table 1 to see if this figure is a curve of constant width.

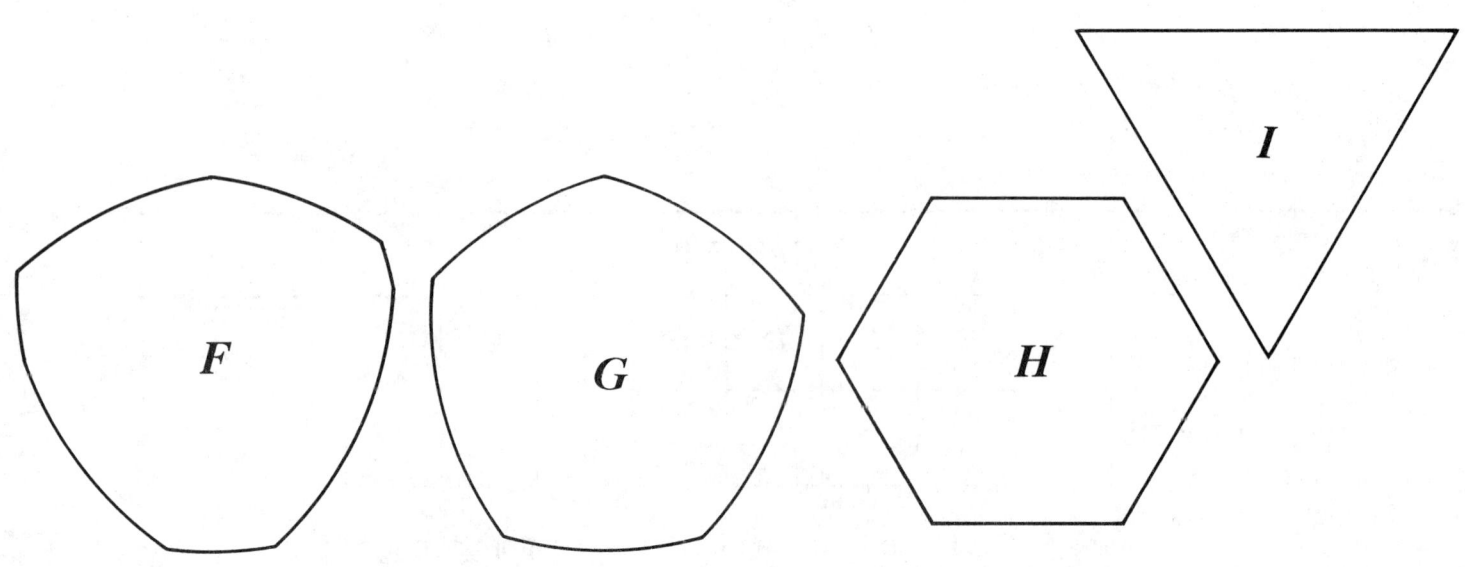

Teaching with *Student Math Notes:* Volume 2

Using the same method, can you construct a curve of constant width on the square, hexagon, and heptagon shown?

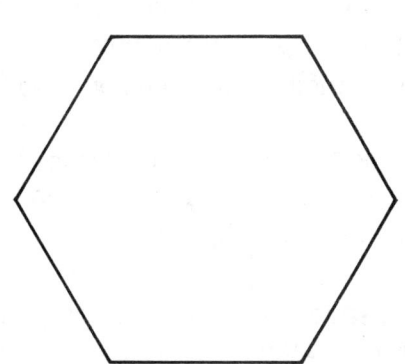
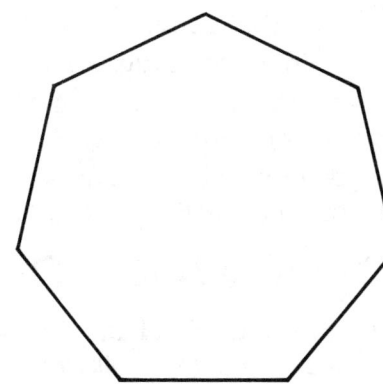

5. Can you construct curves of constant width on any regular polygon using this method? _____ Why or why not? _____

_____

6. What are the restrictions on the number of sides that the polygon can have?

_____

7. Must the generating polygon be regular? _____

Extensions

In the equilateral triangle ABC shown, the sides have been extended an equal distance p. Therefore $AD = AE = BF = BG = CH = CI = p$. Let $AH = q$.

8. What is the relationship among AH, AG, BD, BI, CE, and CF? _____

Open the compass to length p. Place the point of the compass at A and draw $\widehat{DE}$. Place the point at B and draw $\widehat{FG}$, then at C and draw $\widehat{HI}$. Now open the compass to length q. Place the point at A and draw $\widehat{GH}$, at B and draw $\widehat{DI}$, and at C and draw $\widehat{EF}$.

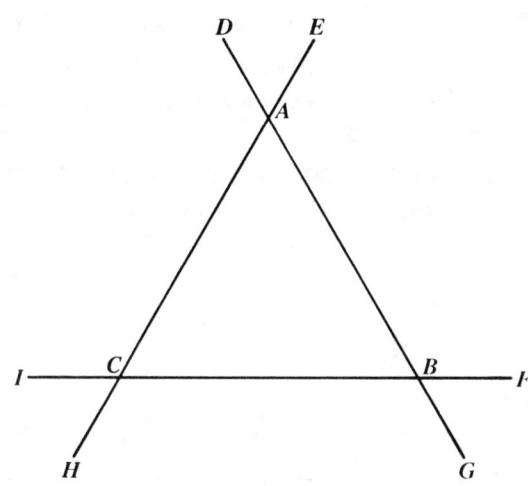

9. Is this new figure a curve of constant width? _____

Trace the pentagon on page 2. Draw the diagonals and extend each one some length p as done on the equilateral triangle. Complete a similar construction for the pentagon.

10. Is the new figure a curve of constant width? _____

11. How can you check your answer? _____

Teaching with *Student Math Notes:* Volume 2

Can You...

- find the perimeter of a circle of width *h*?
- find the perimeter of a Reuleaux triangle of width *h*?
- find the perimeter of a Reuleaux pentagon of width *h*?
- find what appears to be the perimeter of any curve of constant width *h*?
- write a Logo program to generate curves of constant width?
- start with an equilateral triangle, use the method of extension on page 3, and construct a circle?
- determine how the area of the Reuleaux triangle of width *h* compares to a circle of width *h*? (*Hint:* Consider the area of the shaded regions.)

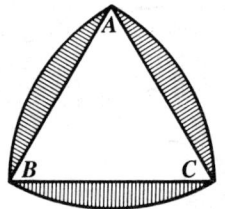

Did You Know That...

- coins in England and Canada have the shape of Reuleaux heptagons? They can be used in vending machines because of their rolling properties as well as their constant width. People in England have complained that the corners cause holes in their pockets.
- the Reuleaux triangle is the basis of a rotary drill that drills square holes? It was invented in 1914 by James Watt and patented in 1917. Note that the vertex goes into the corner (see fig. below), as opposed to the action of a circular bit. With the Reuleaux triangle the motion is eccentric rather than circular. Rotary drills in the shape of pentagons, hexagons, and octagons are also available.

   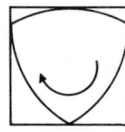

- the rotors in a Wankel rotary engine have the shape of Reuleaux triangles?
- an infinite number of solids have constant width?
- Martin Gardner, famous for his "Mathematical Games" column in *Scientific American,* once said that curves of constant width "provide a sterling example of how far one's mathematical intuition can go astray"?
- curves of constant width were studied by Euler in 1778?
- Reuleaux triangles were named after Franz Reuleaux (1829–1905), an engineer and mathematician who taught at the Royal Technical High School in Berlin, Germany? He was the first person to demonstrate their constant-width properties.
- circles have constant radii and constant diameters? All other curves of constant width have differing "radii" and yet constant diameters (width).

NCTM STUDENT MATH NOTES is published as part of the NEWS BULLETIN by the National Council of Teachers of Mathematics, 1906 Association Drive, Reston, VA 22091. The five issues a year appear in September, November, January, March, and May. Pages may be reproduced for classroom use without permission.

Editor: **Dan Dolan,** Office of Public Instruction, Helena, MT 59620
Editorial Panel: **Johnny W. Lott,** University of Montana, Missoula, MT 59812
Judy Olson, Western Illinois University, Macomb, IL 61455
John G. Van Beynen, Northern Michigan University, Marquette, MI 49855
Editorial Coordinator: **Joan Armistead**
Production Assistants: **Ann M. Butterfield, Sheila C. Gorg**

Printed in U.S.A.

Teacher Notes

Page 81. This is a hands-on activity. Students may choose to predict which of the nine curves have a constant width, but they also need to physically test them using the two lines L_1 and L_2. To have the property of constant width, the curves must be measured in all possible positions, turning them completely around between the lines. However, with the exception of the circle, the location of the pivot point for turning is constantly changing, not fixed as an axle is on a wheel.

Page 82. The constant-width curves constructed here are formed from arcs of a circle and require the use of a compass. An important idea to keep in mind is that all radii for arcs of a given circle are the same length, always measured in a perpendicular fashion to points of tangency on the arcs.

The Reuleaux pentagon will roll along a straight line and maintain a constant height, but there will be no fixed center point as it turns.

Page 83. Students gain a better sense of the nature of curves of constant width when they try to construct them on even-sided polygons such as a square. In the extension figure, notice that the constant curve still consists of arcs of circles, but unlike the earlier constructions, these use arcs from circles with two different radii. The total outside width is always the sum of the lengths of the two different radii.

Page 86. See if students can discover for themselves that two of the curves of constant width come from overlapping three sectors of a given size.

Answers

Page 81. Table 1:

A	B	C	D	E	F	G	H	I
Yes	Yes	No	No	Yes	Yes	Yes	No	No

Page 82. 1. h 2. h 3. h
4. Equal; congruent parts of congruent triangles

Page 83. 5. No; cannot draw two congruent diagonals from the same vertex in even-sided polygons
6. Figures must have an odd number of sides.
7. No   8. Equal   9. Yes   10. Yes
11. Draw two parallel lines L_3 and L_4 that are separated by a distance of $p + q$ and rotate the figure between them as on page 81.

Page 86 (Extension). 1. *a)* Overlap three of the smaller pieces.

b) Overlap three of the larger pieces.

c) Assemble the six remaining pieces.

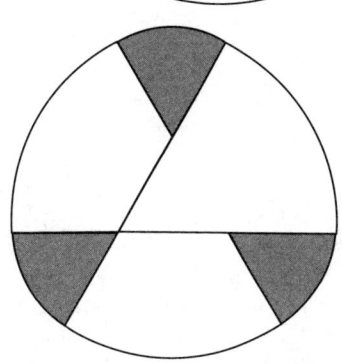

2. *a)* 9.42 cm   *b)* 18.84 cm   *c)* 28.26 cm
3. *a)* 3 cm   *b)* 6 cm   *c)* 9 cm

Teaching with *Student Math Notes*: Volume 2

Extension—CUTOUTS FOR CURVES OF CONSTANT WIDTH

Cut out the two circles and separate each one into six sections by cutting on the marked diameters.

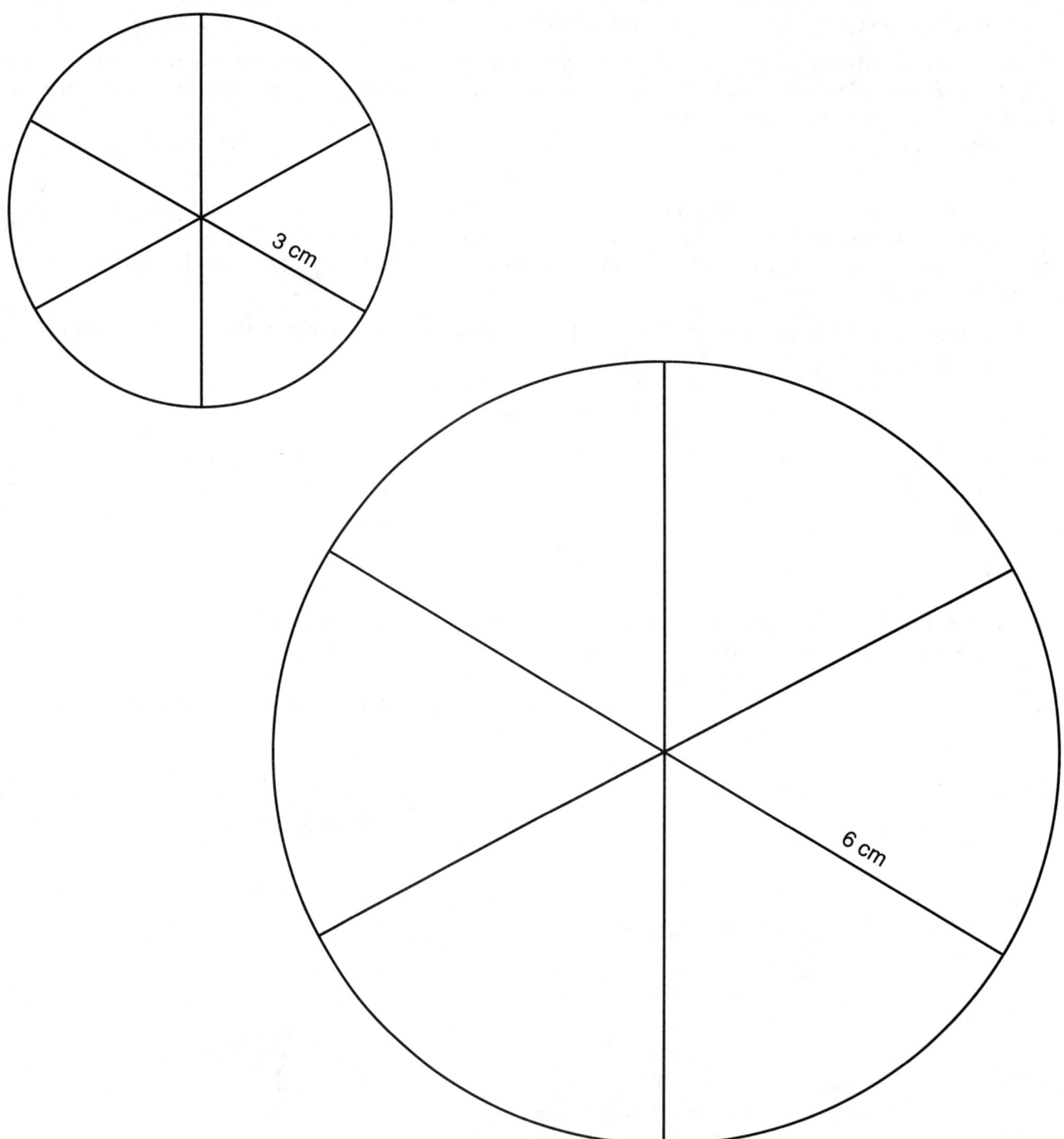

1. Use all twelve pieces from the two circles at one time to form three different curves of constant width.

2. Find the distance around each of the three curves of constant width.

3. Find the constant width of each of the three curves.

Ciphers

Cryptology, the science of writing and breaking codes and ciphers, uses many ideas from mathematics. It is a rapidly expanding profession. Can you decipher the following message, which is written in pig-pen cipher?

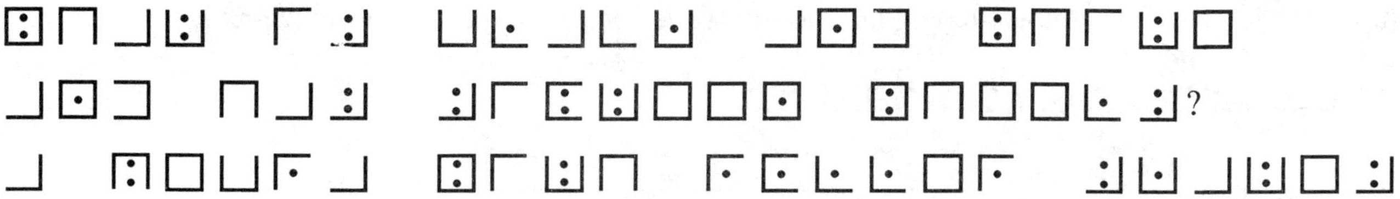

A key to this cipher is formed by placing the letters of the alphabet in "pens" formed by the spaces in a tic-tac-toe grid. In the grid the letters are placed as shown in figure 1.

Fig. 1

A	B	C		J	K	L		S	T	U
D	E	F		M	N	O		V	W	X
G	H	I		P	Q	R		Y	Z	

To make the cipher, no dots are used for the letters A through I, only the sides of the pens. We use one dot in each pen for the letters J through R and two dots per pen for S through Z, as seen in figure 2.

Fig. 2

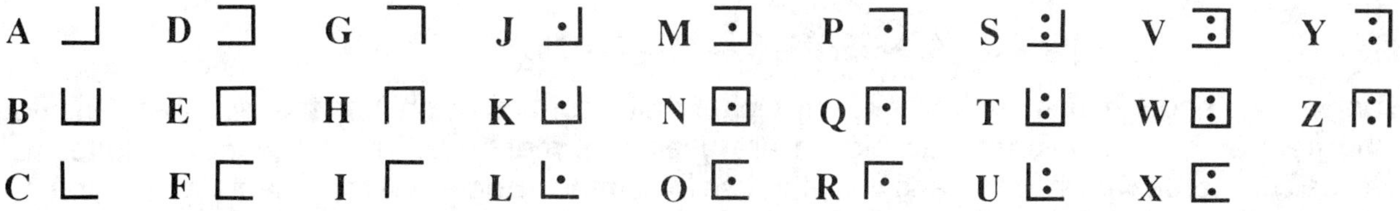

1. Use the key in figure 2 to decipher the message. _____

2. Use the key in figure 2 to encipher the following question: What is red and white on the outside and gray and white on the inside? Answer: A can of Campbell's Elephant Noodle Soup.

The editors wish to thank Anne Teppo, Department of Mathematical Sciences, Montana State University, Bozeman, MT 59715, for writing this issue of *NCTM Student Math Notes*.

Alphabet Shift Cipher

A different type of cipher is formed by shifting the letters in the alphabet by three letters and by wrapping around the alphabet to fill in all twenty-six positions. In the example below, D in plain text is paired with A in cipher text, E is paired with B, and so on.

Cipher text	X	Y	Z	A	B	C	D	E	F	G	H	I	J	K	L	M	N	O	P	Q	R	S	T	U	V	W
Plain text	A	B	C	D	E	F	G	H	I	J	K	L	M	N	O	P	Q	R	S	T	U	V	W	X	Y	Z

To decipher the mystery message shown below, replace each letter given in the cipher line with the letter from the plain text. Space is provided below each line in the message for you to write your answer.

3. TEXQ'P QEB QLRDEBPQ GLY CLO X IFLK?

   _____

   QOVFKD QL CFKA X ABKQFPQ TEL TFII PBB EFJ QTFZB X VBXO

   _____

4. Use the alphabet shift cipher above to encipher the following:

   What is purple and lies at the bottom of the sea?

   _____

   Answer: Moby Grape _____

Cipher Breaking

If the message is long enough, one can "break" any cipher that is formed by shifting the alphabet. One does so by observing the frequency with which certain letters are used in a message. The frequency of letter use in the message can be compared with the frequency of letter use in common English. A representative frequency of letter use can be obtained from the Preamble to the Constitution of the United States.

We the people of the United States, in order to form a more perfect union, establish justice, insure domestic tranquillity, provide for the common defense, promote the general welfare, and secure the blessings of liberty to ourselves and our posterity, do ordain and establish this Constitution of the United States of America.

To determine the frequency of each letter used, we form a table as shown below. The letters A and B are counted for you.

5.

A	B	C	D	E	F	G	H	I	J	K	L	M	N	O	P	Q	R	S	T	U	V	W	X	Y	Z

By observing the frequency of the letters in the message in number 5, one can compare the frequencies of an unknown cipher and decide how the shift was made. In the enciphered message below, determine the frequency of the letters used by filling in the frequency table in number 6. Place this table for the enciphered message under the table in number 5 and shift it until the frequency patterns "match." They will probably not be exact, but you should be able to locate where the shifted cipher alphabet begins. Remember that the cipher wraps around as on page 2.

PAR BL TG XEXIATGM ZKTR, ETKZX, TGW PKBGDEXW? UXVTNLX BY AX

PXKX LFTEE, PABMX, TGW KHNGW, AX PHNEW UX TG TLIBKBG

6. |A|B|C|D|E|F|G|H|I|J|K|L|M|N|O|P|Q|R|S|T|U|V|W|X|Y|Z|

7. How large a shift was made? _____

8. What does the message above say? Write the deciphered message directly below the cipher on the lines provided.

Hidden Word Ciphers

A cipher that is harder to break than the shifted cipher can be formed by using a hidden word in the cipher alphabet and then filling in the rest of the positions with the remaining letters of the alphabet. *Winter* is the hidden word in the following example.

Cipher text	Y|Z|W|I|N|T|E|R|A|B|C|D|F|G|H|J|K|L|M|O|P|Q|S|U|V|X
Plain text	A|B|C|D|E|F|G|H|I|J|K|L|M|N|O|P|Q|R|S|T|U|V|W|X|Y|Z

9. Use the key above to decipher the following. Write the message in the space below the cipher.

SRYO'M ELIV HG ORN AGMAIN YGI WDNYL HG ORN HPOMAIN?

YG NDNJRYGO AG Y ZYEEAN

Teaching with *Student Math Notes:* Volume 2

Can you...

- make up your own cipher, using different symbols for each letter?
- make ciphers harder to break by grouping all the letters in the message into groups of four or five? For example, TJPV MZNH VMOO JADB PMZO CDNJ PO.

Did you know that...

- a code is a system that substitutes symbols, words, or groups of letters for phrases and words, whereas a cipher is a system that substitutes symbols or letters for each individual letter?
- the Morse code is really a cipher?
- Navahos were employed to send messages by telephone during World War II because the Navaho language served as a natural "code" that could be sent and received by native speakers, but was impossible to understand by any non-Navaho-speaking listener?
- the following list contains stories that involve some kind of cipher:

  "The Gold Bug" by Edgar Allan Poe

  "The Dancing Men," a Sherlock Holmes story by A. Conan Doyle

  "The First Letter" in *Just So Stories* by Rudyard Kipling

  The Chain of Death by Maxwell Grant

For more information...

- Gardner, Martin. *Codes, Ciphers and Secret Writing.* New York: Simon & Schuster, 1972.
- Jacobs, Harold. *Mathematics: A Human Endeavor.* San Francisco: W. H. Freeman & Co., 1970.
- Peck, Lyman C. *Secret Codes, Remainder Arithmetic and Matrices.* Reston, Va.: National Council of Teachers of Mathematics, 1975.
- Sinkov, Abraham. *Elementary Cryptanalysis—a Mathematical Approach.* Washington, D.C.: Mathematical Association of America, 1966.
- Snow, Joanne R. "An Application of Number Theory to Cryptology." *Mathematics Teacher* 82 (January 1989): 18–26.
- Zim, Herbert S. *Codes and Secret Writing.* New York: William Morrow & Co., 1948.

NCTM STUDENT MATH NOTES is published as part of the **NEWS BULLETIN** by the National Council of Teachers of Mathematics, 1906 Association Drive, Reston, VA 22091. The five issues a year appear in September, November, January, March, and May. Pages may be reproduced for classroom use without permission.

Editor: **Dan Dolan,** Office of Public Instruction, Helena, MT 59620
Editorial Panel: **Johnny W. Lott,** University of Montana, Missoula, MT 59812
Judy Olson, Western Illinois University, Macomb, IL 61455
John G. Van Beynen, Northern Michigan University, Marquette, MI 49855
Editorial Coordinator: **Joan Armistead**
Production Assistants: **Ann M. Butterfield, Sheila C. Gorg**

Printed in U.S.A.

Teacher Notes

Page 87. A cipher is a method for changing text into a secret version. To encipher is to apply the cipher to the text to generate the secret version. To decipher is to translate back into the original text. The cipher on this page changes letters into geometric figures. The segments locate the letters' positions in a 3 × 3 array. The number of dots indicates which of the three arrays is used. Students may choose to create a similar cipher by placing each 3 × 3 array of letters in a square and using segments as they are located on the outside boundaries.

Page 88. The shift cipher given here associates every letter in the original plain text with the corresponding letter three places before it in the cipher text. To encipher, move three letters back. To decipher, move three letters ahead. In each case, use a wraparound alphabet to fill all positions.

Breaking a shift cipher requires finding how many letters are skipped in the shift. Clearly, it can be anything from one to twenty-five letters. A statistical analysis of the frequency of the letters used in the cipher text often reveals the magnitude of the shift. Have students compare the frequencies of letter use found in the Preamble to the United States Constitution with other written passages of approximately the same length.

Breaking any substitution cipher can be much harder because letters can be matched in 26!, or more than 400 000 000 000 000 000 000 000 000, different ways.

Page 89. The cipher at the bottom of the page contains the hidden word *winter*. Have students find how many different ciphers of this type could be formed using the word *winter* and whether or not the words *spring, summer,* and *fall* could also be used.

Page 92. To facilitate student understanding of the process, you may want to begin by constructing another coded spirolateral design with the class. Three basic sets of designs will emerge through this process, two- and four-part shapes that come back on themselves and those that spin off to infinity. Encourage students to discover what properties of the name determine which of these behaviors arises.

In the example for Euler, the repeating four-step direction cycle RDLU is matched with the repeating five-step number cycle 53359. Five-direction cycles fit together with four-number cycles, bringing the path back to the starting point. Stress the clockwise turning from step to step, both within the number sequence and when the sequence repeats itself. On the first cycle, the 5 starts to the right; on the second, it starts down; on the third, it starts to the left; and on the fourth, it starts up. The transformation applied to the geometric path to move it from one cycle to the next in the spirolateral can be described through a rotation or turn and a translation or slide.

The behavior for the name Euclid in question 1 is different because the coded number cycle contains a six-digit sequence. Here, three-direction cycles fit together with two-number cycles, bringing the path back to the starting point more quickly and yielding a spirolateral with a somewhat different appearance.

Answers

Page 87.

Page 88. 3. What's the toughest job for a lion? Trying to find a dentist who will see him twice a year.

4. TEXQ FP MROMIB XKA IFBP XQ QEB YLQQLJ LC QEB PBX? JLYV DOXMB

5.

A	B	C	D	E	F	G	H	I	J	K	L	M	N	O	P	Q	R	S	T	U	V	W	X	Y	Z
14	4	7	11	39	9	2	9	20	1	0	10	7	17	25	6	1	19	21	29	10	2	2	0	3	0

Page 89. 6.

A	B	C	D	E	F	G	H	I	J	K	L	M	N	O	P	Q	R	S	T	U	V	W	X	Y	Z
5	6	0	1	6	1	8	2	2	0	6	4	2	3	0	5	0	2	0	10	2	1	5	12	0	2

7. A shift of seven
8. Why is an elephant gray, large, and wrinkled? Because if he were small, white, and round, he would be an aspirin.
9. What's gray on the inside and clear on the outside? An elephant in a Baggie

Page 93. 1. Euclid: 533394

Page 94. 2. Pascal: 711313
Pythagoras: 7728176911
Math: 4128

The three spirolaterals for the mathematicians contain cycles that repeat only twice before returning to their starting point. However, the coded number sequence for MATH has just four digits. Its spirolateral spirals off to the upper right, repeating endlessly, never returning to the starting point.

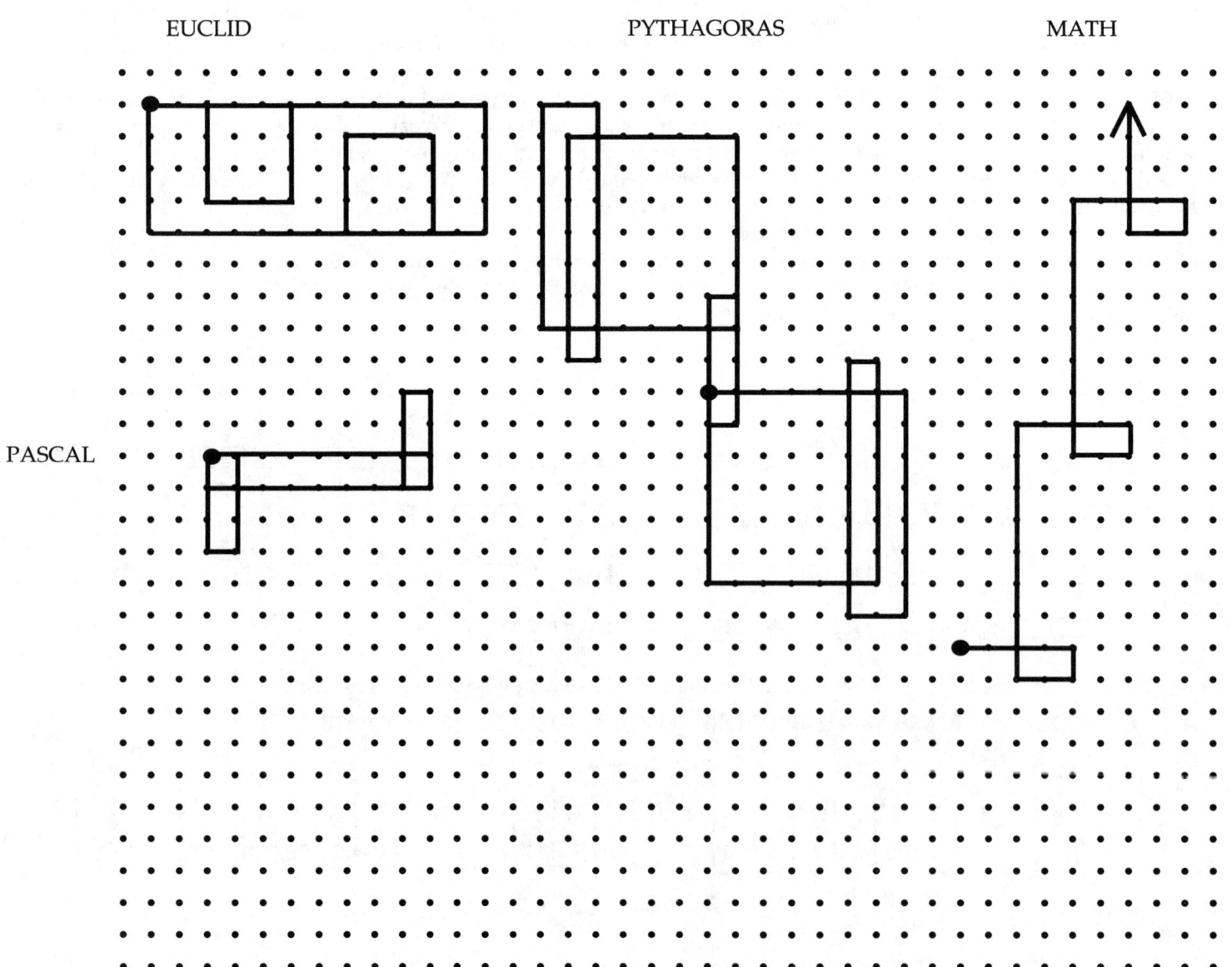

Extension—CODED NAMES IN DESIGNS

Euler was a famous Swiss mathematician (1707–83) who was interested in networks and paths. Using this key, his name would appear as 53359.

A	B	C	D	E	F	G	H	I	J	K	L	M	N	O	P	Q	R	S	T	U	V	W	X	Y	Z
1	2	3	4	5	6	7	8	9	1	2	3	4	5	6	7	8	9	1	2	3	4	5	6	7	8

Spirolaterals are geometric designs formed by repeating sequences of geometric steps on a grid. In this case, move clockwise on a square grid, turning 90° at each step.

Right   Down   Left   Up   R   D   L   U

Now repeat endlessly the sequence of four-letter turns and the coded five-digit sequence for the name Euler, matching letters with numbers.

R	D	L	U	R	D	L	U	R	D	L	U	R	D	L	U	R	D	L	U	...
5	3	3	5	9	5	3	3	5	9	5	3	3	5	9	5	3	3	5	9	...

Finally, generate a design showing this pairing as steps on a grid. Notice that the path returns to the starting point, producing a coded geometric design for the name Euler.

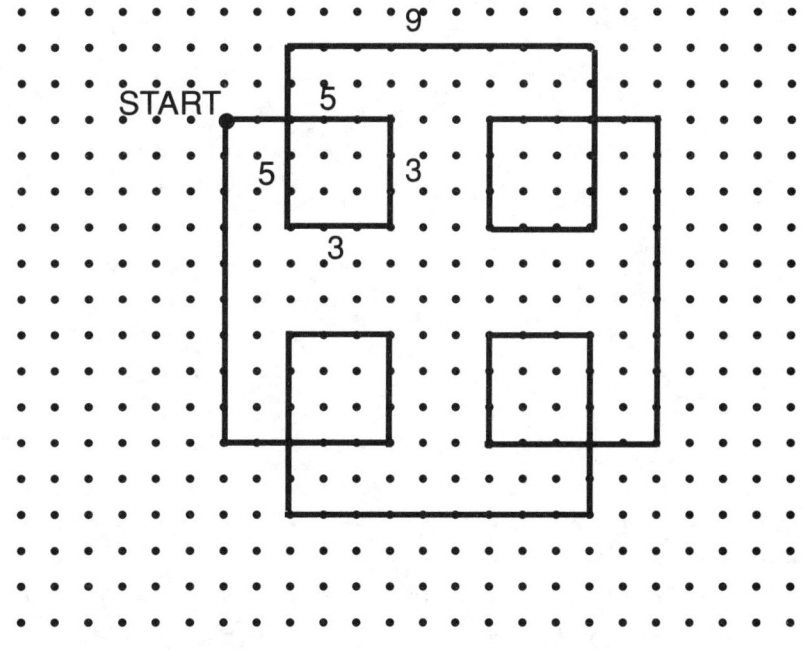

1. Give the numerical code for the ancient Greek mathematician Euclid. Then use the code to draw the corresponding coded spirolateral design.

Teaching with *Student Math Notes*: Volume 2

2. Use this grid to draw the coded spirolaterals for the French mathematician Pascal (1623–62), and the ancient Greek mathematician Pythagoras (540 B.C.). Then try drawing the coded spirolateral for the word *math*. Compare and comment on your results.

NATIONAL COUNCIL OF TEACHERS OF MATHEMATICS   SEPTEMBER 1989

QUADRI-GON(E)?

Tanya had read "Midpoint Madness" in the January 1988 issue of *Student Math Notes* and wanted to design a quadrilateral *WXYZ* having the vertices of parallelogram *ABCD* as midpoints of its four sides. If *A* is the midpoint of $\overline{WX}$, *B* is the midpoint of $\overline{XY}$, *C* is the midpoint of $\overline{YZ}$, and *D* is the midpoint of $\overline{ZW}$, use figure 1 to help her design the quadrilateral.

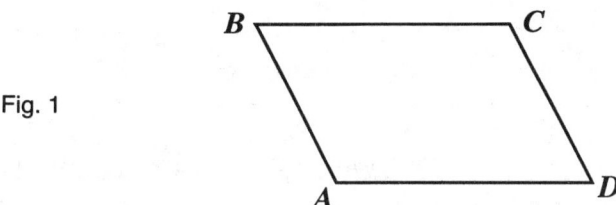

Fig. 1

In figure 2a–d, use point *W* as one vertex of the desired figure, which has points *A*, *B*, *C*, and *D* as midpoints as in figure 1.

1. Find the other vertices, *X*, *Y*, and *Z*, and draw figure *WXYZ*.

Fig. 2

The editors wish to thank Gerald White and Melfried Olson, Western Illinois University, Macomb, IL 61455, and Tanya Olson, Macomb High School, Macomb, IL 61455, for writing this issue of *NCTM Student Math Notes*.

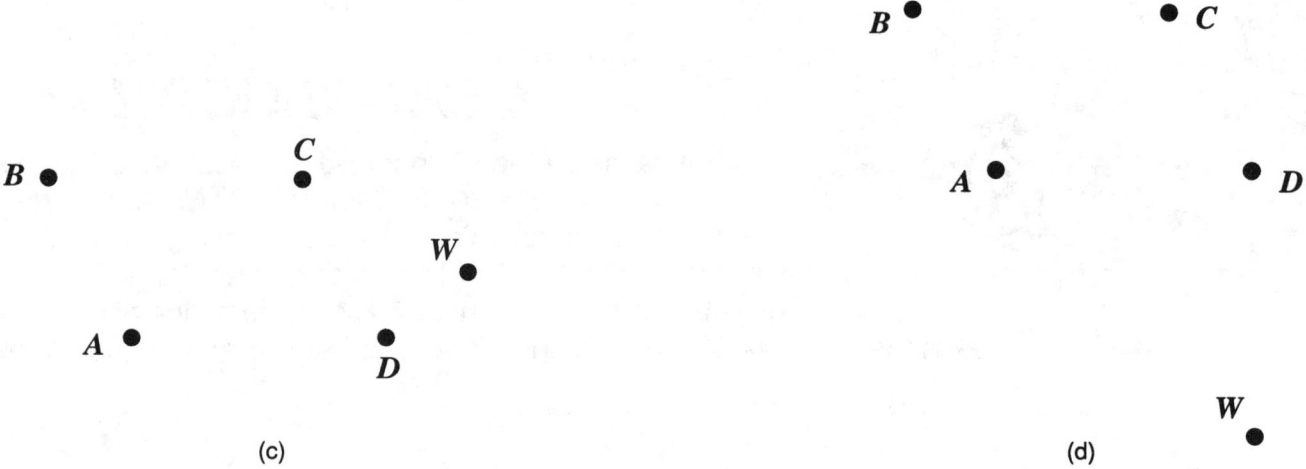

(c) (d)

Fig. 2—Continued

We call the types of designs found in figure 2 *quadrigons*. Figure 2a shows a convex quadrigon; figures 2b and 2c show cross quadrigons; and figure 2d shows a concave quadrigon. In figure 3, locate points W, X, Y, and Z to draw the specified quadrigons.

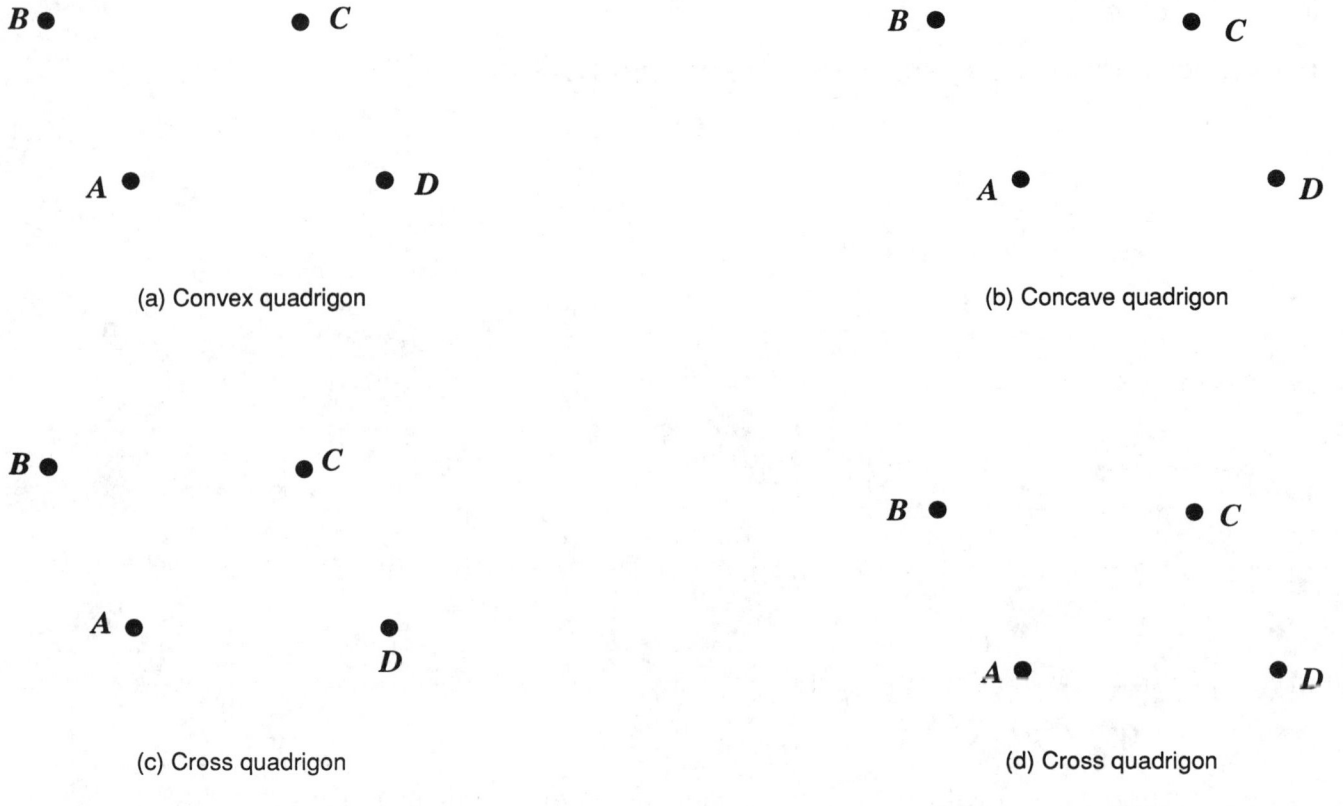

(a) Convex quadrigon (b) Concave quadrigon

(c) Cross quadrigon (d) Cross quadrigon

Fig. 3

In figure 4, a grid is used to change the focus from an infinite number of points to the same problem focused on a discrete set of regions.

2. Using figure 4, can you locate in the same region two points, W_1 and W_2, that yield different kinds of quadrigons?

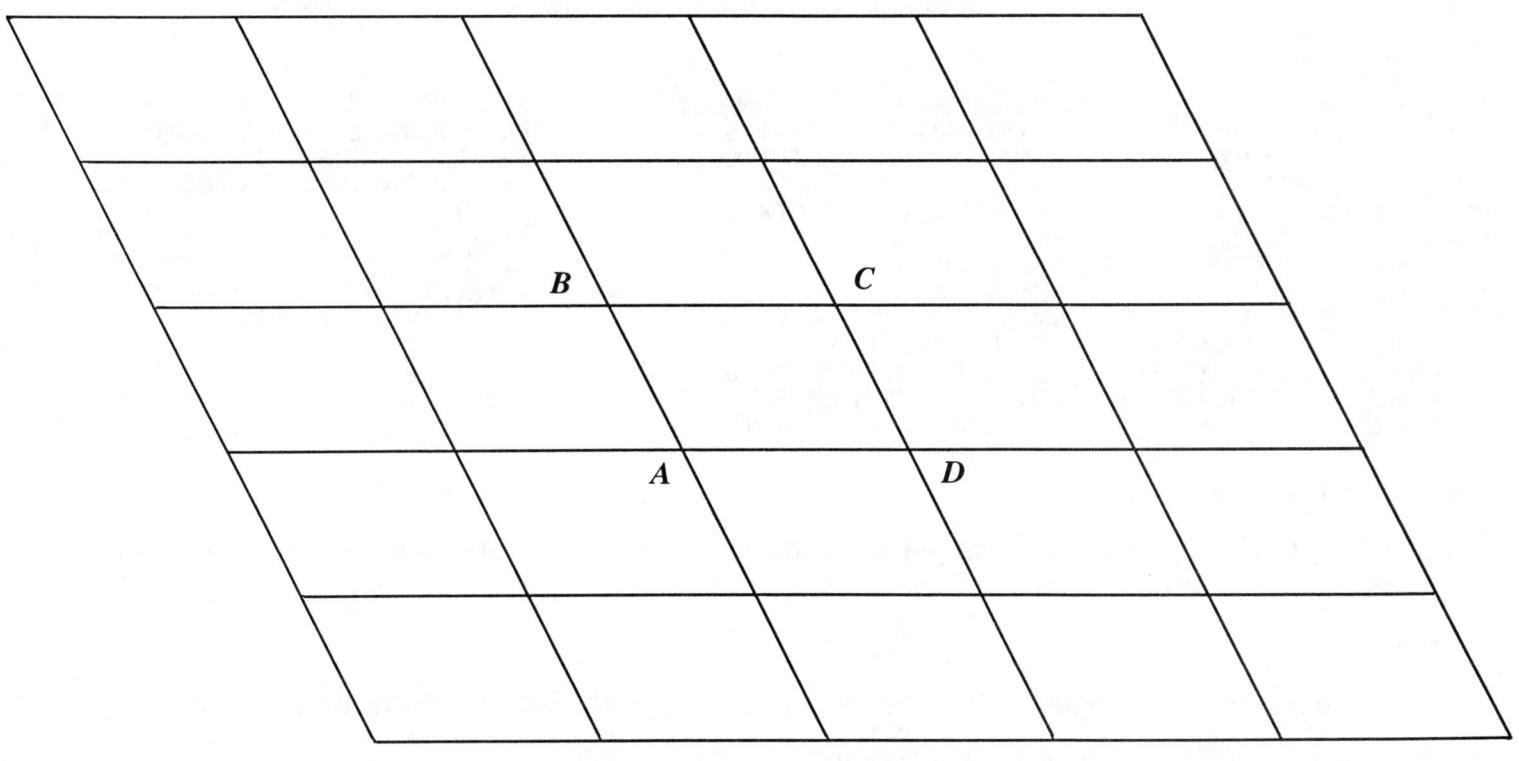

Fig. 4

In figure 4, use yellow to color regions that generate concave quadrigons, pink to color regions that generate cross quadrigons, and leave white the regions that generate convex quadrigons.

3. What type of quadrigon is generated using points on the boundaries between white and yellow regions? _____

4. What type of quadrigon is generated using points on the boundaries between white and pink regions? _____

5. What type of quadrigon is generated using points on the boundaries between pink and yellow regions? _____

The quadrigons found using points on the boundaries are either triangle quadrigons or flag quadrigons. In figure 4, use blue to color the boundaries generating triangle quadrigons and red to color the boundaries generating flag quadrigons.

Computer Investigation

The following computer programs, written in MIT Terrapin Logo, can be used to investigate quadrigons in a dynamic manner. If Logo is available, enter the following procedures in the computer and type STARTUP to begin the investigation. When STARTUP is executed, the monitor screen is divided into regions. Use FORWARD, BACK, RIGHT, or LEFT to move the turtle to a point in one of the regions. (No trail is drawn.) Next execute QUADRIGON and observe what type of quadrigon is drawn. Then move the turtle along a line parallel to one of the boundaries to another point in the region and execute QUADRIGON again. Continue this process to determine what types of quadrigons are generated from each region. If at any time the screen becomes too cluttered, clear the screen and start over.

```
TO STARTUP
PENDOWN RIGHT 50
FD 100 BK 200 FD 100
FD 20 MAKE "BX XCOR
MAKE "BY YCOR RT 40
FD 100 BK 200 FD 100
FD 30 MAKE "CX XCOR
MAKE "CY YCOR RT 140
FD 100 BK 200 FD 100
FD 20 RT 40 FD 100 BK 200
FD 100 FD 30 RT 140 RT 180
PC 3 REPEAT 2[FD 20 LT 140
   FD 30 LT 40]
PC 1 SETH 0 PU PRINT [WHEN
   THIS PROCEDURE ENDS,
   MOVE THE CURSOR TO
   VERTEX W, AND THEN
   EXECUTE QUADRIGON.]
END
TO QUADRIGON
PENDOWN
MAKE "WX XCOR MAKE "WY YCOR
SETH TOWARDS 0 0
FD 2*DISTANCE :WX :WY 0 0
SETH TOWARDS :BX :BY
FD 2*DISTANCE XCOR YCOR :BX :BY
SETH TOWARDS :CX :CY
FD 2*DISTANCE XCOR YCOR :CX :CY
SETH TOWARDS 30 0
FD 2*DISTANCE XCOR YCOR 30 0
PU HT
END
TO DISTANCE :X1 :Y1 :X2 :Y2
OP SQRT ((:X1-:X2)*(:X1-:X2)+
   (:Y1-:Y2)*(:Y1-:Y2))
END
```

Did you know that . . .

- point W need not be on the same plane as the four points to create a quadrigon? (What type of quadrilateral is formed if point W is not in the plane of points A, B, C, and D?)

Can you . . .

- prove that the area of the figure determined by the four midpoints is one-half the area of the quadrigon?
- determine if the pattern found in the previous question continues?
- find quadrigons that are rectangles? Parallelograms? Trapezoids?

Teacher Notes

Page 95. The January 1988 issue of *Student Math Notes* found on pages 45–50 explored the polygons formed by connecting the midpoints of given quadrilaterals. This activity does the reverse, exploring the nature of the original quadrilaterals given the midpoints of their sides. In each case, students need to study the initial points carefully, being sure that *A*, *B*, *C*, and *D* are used as the midpoints and *W* as a single vertex of the original quadrilateral. Encourage careful and accurate measuring and constructions.

Page 96. Clearly, as the location of the vertex *W* moves from one position to another, the properties of the corresponding quadrigon change. The three special cases covered are convex, concave, and cross quadrigons. Students will begin to discover the regions where these occur when working on the last four figures on page 96. The complete analysis occurs on page 97.

Page 100. A straw model of the tetrahedrons with the midpoint segments on the faces included can be a very helpful teaching aid here. Use twist ties to connect the straws. In all, there are fifteen possible ways to connect midpoints. The three that are not on the faces of the tetrahedron locate the main diagonals of the resulting regular octahedron. These are also the diagonals of the three squares that lie on the edges of the solid.

Answers

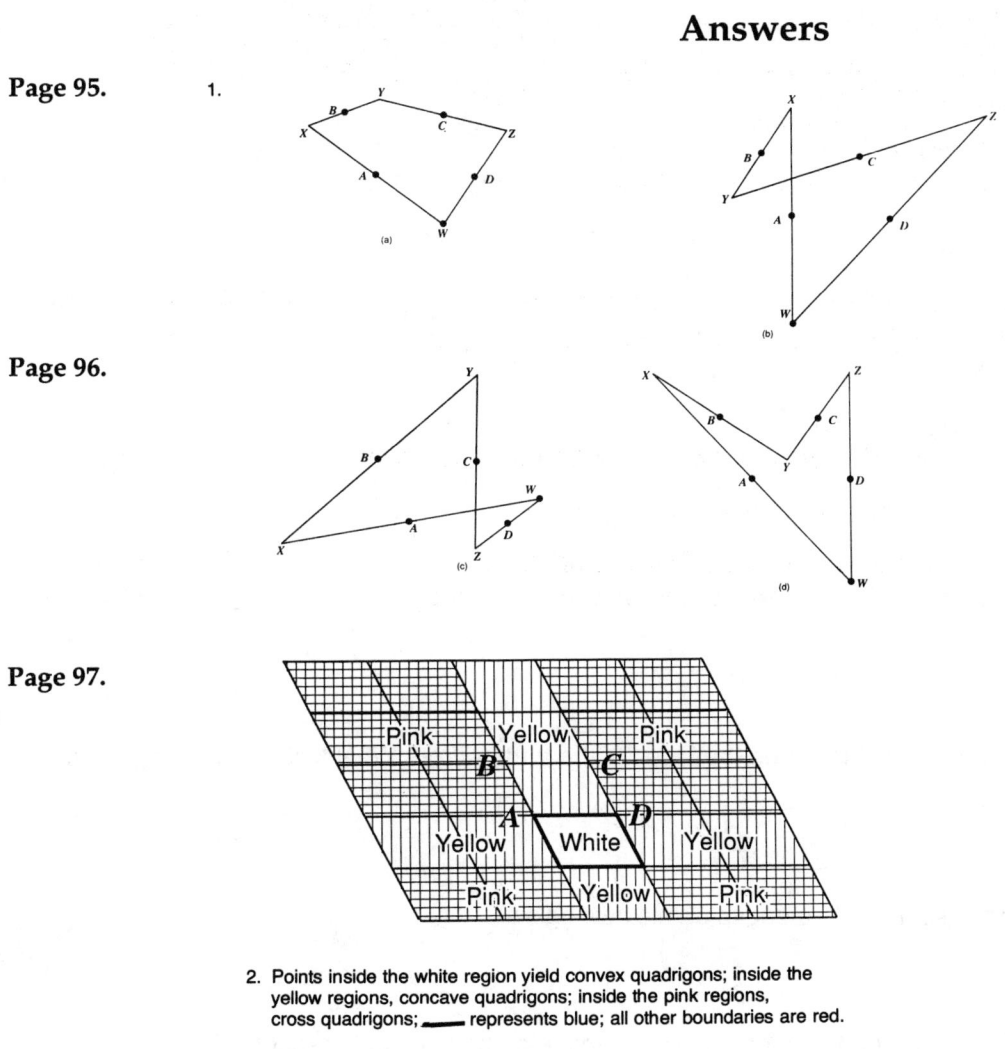

2. Points inside the white region yield convex quadrigons; inside the yellow regions, concave quadrigons; inside the pink regions, cross quadrigons; ——— represents blue; all other boundaries are red.

3. Triangle   4. Triangle   5. Flag

Page 100 (Extension).   **1.** 6; 15   **2.** A regular octahedron   **3.** Eight equilateral triangles and three squares
4. Half the surface area and half the volume of the original tetrahedron

Extension—MORE ON MIDPOINTS

If you connect the midpoints of the sides of a triangle, you get three connected segments that form another triangle similar to the original with one-fourth the perimeter and one-half the area. What do you get if you do the same thing with the edges of a tetrahedron?

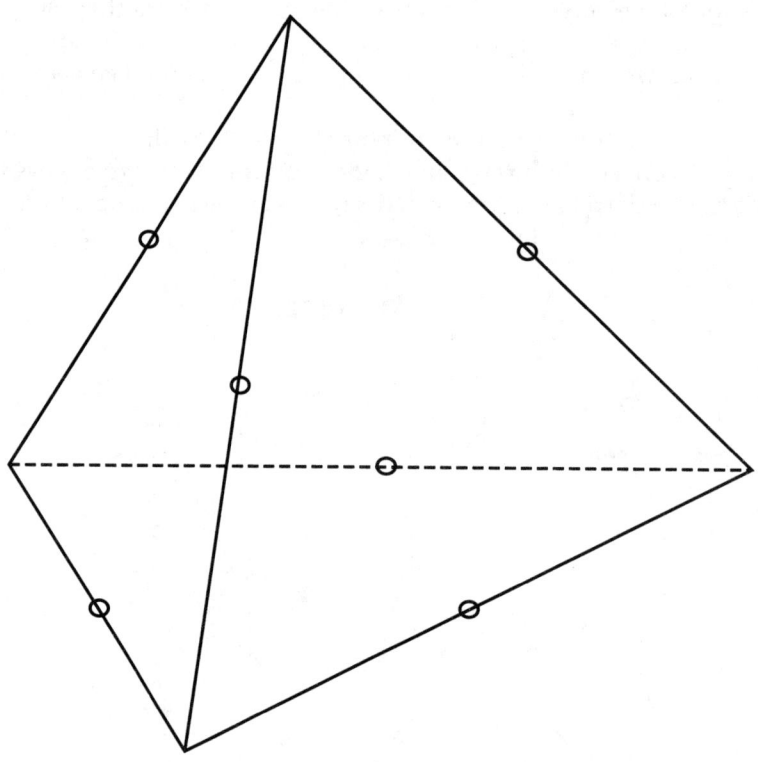

1. How many edges are in the regular tetrahedron shown? How many ways can you connect the midpoints of these segments, two at a time?

2. Draw in the figure only those connecting segments that lie on the faces of the tetrahedron. What kind of figure do these segments form?

3. Name all the polygons that can be formed using only the edges of this new figure.

4. How do the surface area and volume of the resulting figure compare with those of the original tetrahedron?

NATIONAL COUNCIL OF TEACHERS OF MATHEMATICS   NOVEMBER 1989

Godzilla®: Fact or Fiction?

Could Godzilla® really have been tall enough to look into the windows of skyscrapers as movies lead us to believe? Have animals such as Godzilla, King Kong, Rodan, or Mothra ever roamed the earth? The *Tyrannosaurus rex,* which resembles Godzilla®, was the largest meat-eating animal that ever lived and was only about twenty feet tall and forty-five feet long with a four-foot head. Why have fossils of an animal the size of Godzilla® or King Kong never been found?

Similarity

Similarity can be used to explore the possible existence of Godzilla®-sized creatures. Figure 1 shows two similar squares. Similar figures have the same shape. In figure 2, construct a pentagon similar to the given pentagon such that (1) l_2 corresponds to l_1 and is twice as long and (2) each of the other sides is two times the length of its corresponding side. In figure 3, construct a triangle similar to the given triangle such that (1) l_2 corresponds to l_1 and is three times as long and (2) each of the other sides is three times the length of its corresponding side.

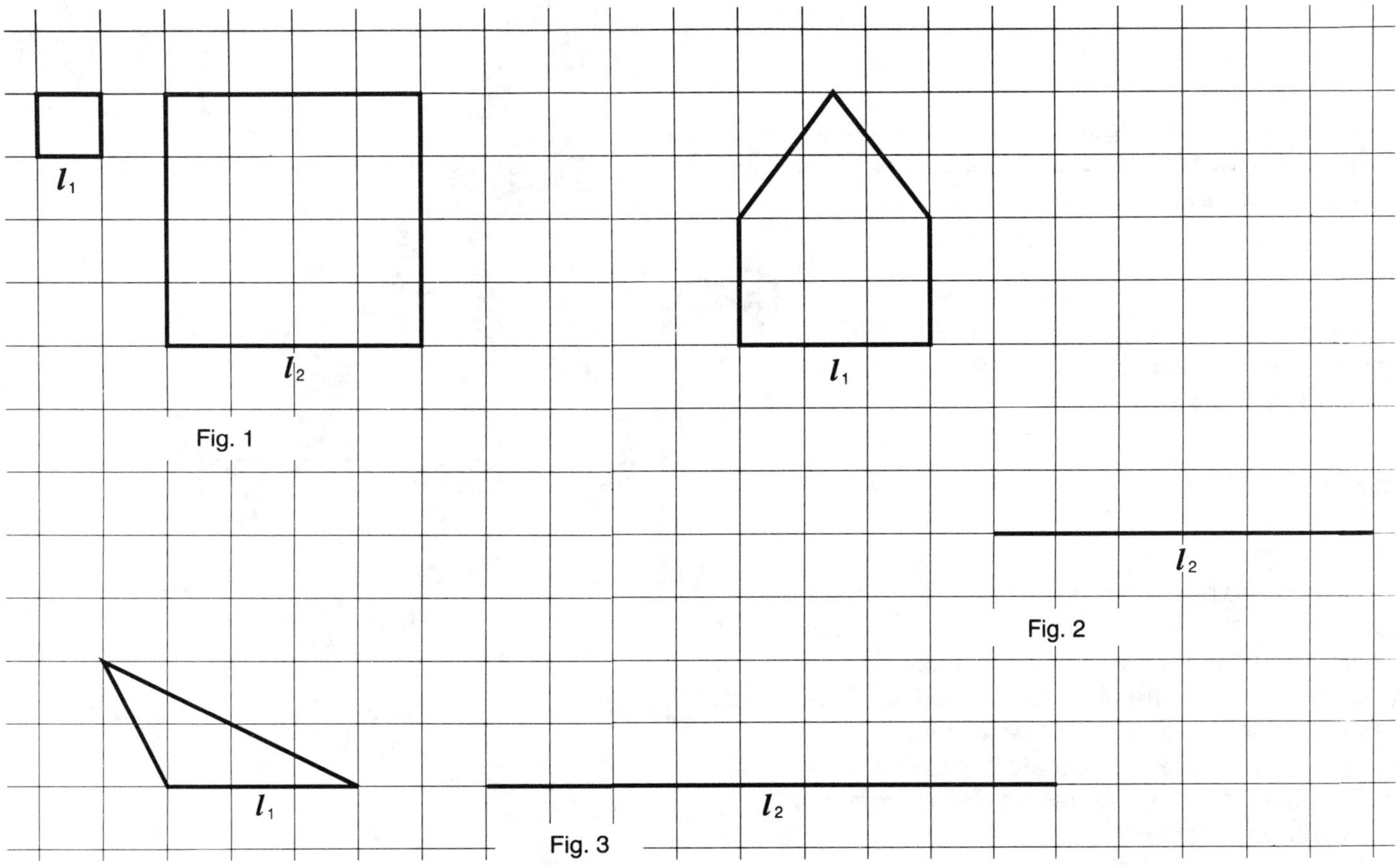

The editors wish to thank Rick Billstein, University of Montana, Missoula, MT 59812, and Jim Trudnowski, Carroll College, Helena, MT 59620, for writing this issue of *NCTM Student Math Notes*.

Write the measurements from figures 1–3 in table 1. From that information, complete table 1. A_1 and A_2 are the areas of the respectively similar figures. Leave all ratios in reduced form.

Table 1

	l_1	l_2	A_1	A_2	l_1/l_2	A_1/A_2
Figure 1						
Figure 2						
Figure 3						
Figure 4						

1. Use the information in table 1 to explain how the ratio of the areas of two similar figures is related to the ratio of the lengths of the corresponding sides. _____

2. If the ratio of the lengths of the corresponding sides of the two similar figures is 2/3, what is the ratio of their areas? _____

3. If the ratio of the lengths of the corresponding parts of two similar creatures is 3/5, what is the ratio of their areas? _____

Choose any two points on the outline of the smaller Godzilla®. Let the length of the line segment connecting the two points be l_1. Find the corresponding points on the outline of the larger Godzilla®. Let the length of this line segment be l_2. Measure the lengths of l_1 and l_2 and estimate the areas of the two Godzilla® figures. Write the measurements and estimates in table 1.

Fig. 4

Three-dimensional Objects

To determine the relative volumes of similar objects, we study cubes of various sizes. In figure 5, l_1 is the length of each face; the volume of the cube, V_1, is one cubic unit. Use $1 \times 1 \times 1$ cubes to construct larger cubes whose sides are two times, three times, and four times as large as the given cube in figure 5. Put the results in table 2.

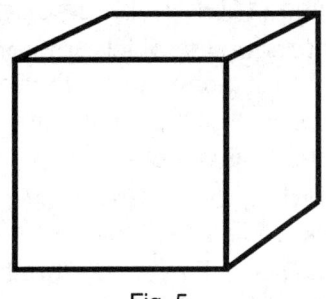

Fig. 5

Table 2

Cube	l_1	l_2	V_1	V_2	l_1/l_2	V_1/V_2
$2 \times 2 \times 2$	1		1			
$3 \times 3 \times 3$	1		1			
$4 \times 4 \times 4$	1		1			
$n \times n \times n$	1		1			

4. Use the information in table 2 to explain how the ratio of the volumes of two cubes is related to the ratio of the lengths of their sides. _____

5. If the ratios of the volumes of three-dimensional representations of Godzilla® follow the generalization you wrote in question 4, use that information and table 1 to find the ratio of the volumes of the two Godzillas® in figure 4. _____

How Tall Could Godzilla® Be?

If we enlarge an animal to a giant of similar proportions by a factor of 20, then height, width, and depth are all multiplied by 20. The animal's weight and volume would each increase by 20^3. We know from John Haldane's *On Being the Right Size in Possible Worlds* (1928) that a human thigh bone can support weight up to only ten times the weight of a six-foot person. In a similar manner, suppose Godzilla's® thigh bone could support weight up to ten times the weight of a twenty-foot animal.

6. If Godzilla® is forty feet tall, twice the height of a similar twenty-foot animal, its weight is how many times greater than the weight of the twenty-foot animal? _____

7. Could Godzilla® be forty feet tall? _____

8. Could Godzilla® be tall enough—possibly 100 feet—to look into the windows on the upper stories of a ten-story building? _____

Explain your answer. _____

Did you know that . . .

- Galileo (1564–1642) predicted that the tallest tree could not exceed 300 feet?
- giant sequoias, which grow only on the West Coast of the United States, have grown as tall as 360 feet? The bark is up to 2 inches thick, the trunk is 30 feet in diameter, and some trees are over 3000 years old?
- a fish, in doubling its length, multiplies its weight by about eight? a fish doubles its weight in growing from four inches to five inches long?
- according to the *Guiness Book of World Records,* the tallest man on record was Robert Wadlow from Alton, Illinois? He was 8 feet, 11.1 inches.
- in *Gulliver's Travels* by Jonathan Swift, Gulliver visited the land of Brobdingnag, where the people were similar to humans but about twelve times as tall?

Can you . . .

- find the weight of a nine-inch-tall Brownie named Franjenn, a character in the movie *Willow*? Assume that he is similar to Madmartigan, a human, who is 6 feet tall and weighs 180 pounds?
- find the weight of a Lilliputian, a character similar to humans, in *Gulliver's Travels*, who is 1/12 the height of Gulliver, who weighs 185 pounds?

References

- Garfunckel, Solomon, project director. *For All Practical Purposes: Introduction to Contemporary Mathematics.* New York: W. H. Freeman Co., 1988.
- Haldane, John B. S. *On Being the Right Size in Possible Worlds.* New York: Books for Libraries Press, 1928.
- Thompson, D'Arcy. *On Growth and Form.* New York: Cambridge University Press, 1961.

Teacher Notes

Page 101. Discuss how the ratio of similarity is found from the ratio of lengths of corresponding sides. Once the figures are drawn, consider the similarity in each pair of figures as a transformation. Locate the point of projection that can be used to show this relationship as a transformation.

Page 102. Again, have students view the two figures as a transformation from one to the other. From a location above and to the right of the smaller figure, just off the page, is a point from which the larger figure can be viewed as a projection of the smaller.

The relationship developed on this page can be extremely useful when studying similar figures, especially those classes of figures that are always similar, such as circles and spheres and squares and cubes. When two figures are similar, the ratio of their areas is always the square of the ratio of similarity.

Page 103. This page develops a very important relationship. When two three-dimensional figures are similar, the ratio of their volumes is always the cube of the ratio of similarity. The book *On Growth and Form*, listed in the references on page 104, is an excellent source of additional information on the questions raised in this activity.

Page 106. This hands-on activity blends a set of similar figures, triangles, with another set, trapezoids, that are not similar. In the first case, successive ratios of areas and perimeters remain fixed because the folded triangular flaps are all similar equilateral triangles. In the second case, they do not remain fixed because the trapezoids are not similar.

Answers

Pages 102 and 103.

Table 1

	l_1	l_2	A_1	A_2	l_1/l_2	A_1/A_2
Figure 1	1	4	1	16	1/4	1/16
Figure 2	3	6	9	36	1/2	1/4
Figure 3	3	9	3	27	1/3	1/9
Figure 4	Will vary				~1/2	~1/4

Table 2

Cube	l_1	l_2	V_1	V_2	l_1/l_2	V_1/V_2
$2 \times 2 \times 2$	1	2	1	8	1/2	1/8
$3 \times 3 \times 3$	1	3	1	27	1/3	1/27
$4 \times 4 \times 4$	1	4	1	64	1/4	1/64
$n \times n \times n$	1	n	1	n^3	$1/n$	$1/n^3$

1. The ratio of the areas is the square of the ratio of the corresponding sides.
2. 4/9
3. 9/25
4. The ratio of the volumes is the cube of the ratio of the corresponding sides.
5. –1/8
6. Eight times
7. It is possible.
8. No. The thigh bones could not support the weight that corresponds to that height, which is approximately 125 times the weight of a twenty-foot animal.

Page 106 (Extension). Successive folds from the same vertex form smaller and smaller triangular flaps.
1. Yes; yes
2. 1:4; 1:2
3. 1:4; 1:2 and 1:16; 1:4
4. 1:256; 1:16
5. No
6. 15:12; 10:11 and 1023:768; 95:80

Extension—FLAP FOLDING

Cut out this equilateral triangle. Fold the marked vertex to the midpoint of the opposite side. Open the triangle and fold the same vertex to the midpoint of the crease formed. Open and again fold the vertex to the midpoint of the new crease. Repeat the process through five steps, each time folding over a smaller triangular flap than the time before.

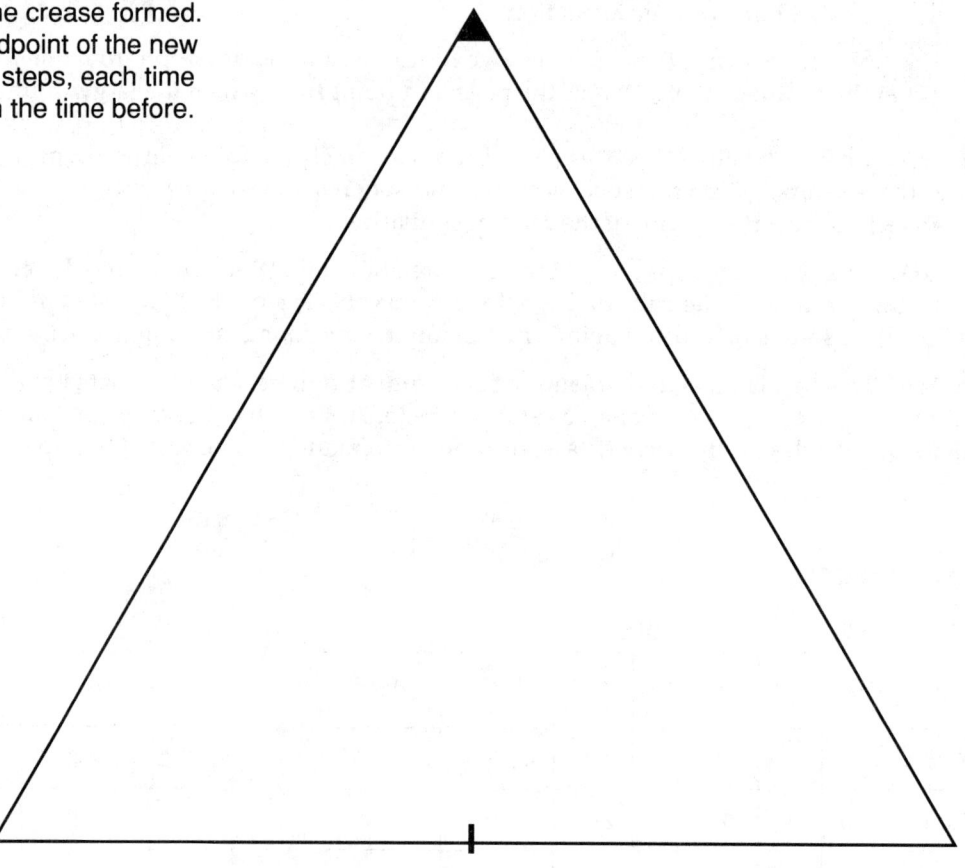

1. Are the five different triangular flaps folded over in the five steps similar to each other? Are they similar to the original equilateral triangle?

2. How do the area and perimeter of the first folded triangular flap compare with those of the original equilateral triangle?

3. How do the area and perimeter of the second folded triangular flap compare with those of the first folded flap? How do they compare with the original equilateral triangle?

4. How do the area and perimeter of the last folded triangular flap compare with those of the first folded flap?

5. Are the trapezoids formed at each step similar to each other?

6. How do the areas and perimeters of the trapezoids in steps 1 and 2 compare? What about those for steps 1 and 5?

5-con Triangles

In figure 1 four triangles have been placed in a spiral pattern. Your challenge is to extend this spiral by adding appropriate triangles to its ends. Carefully cut out the four triangles and label the interior of each angle with the letter of its corresponding vertex.

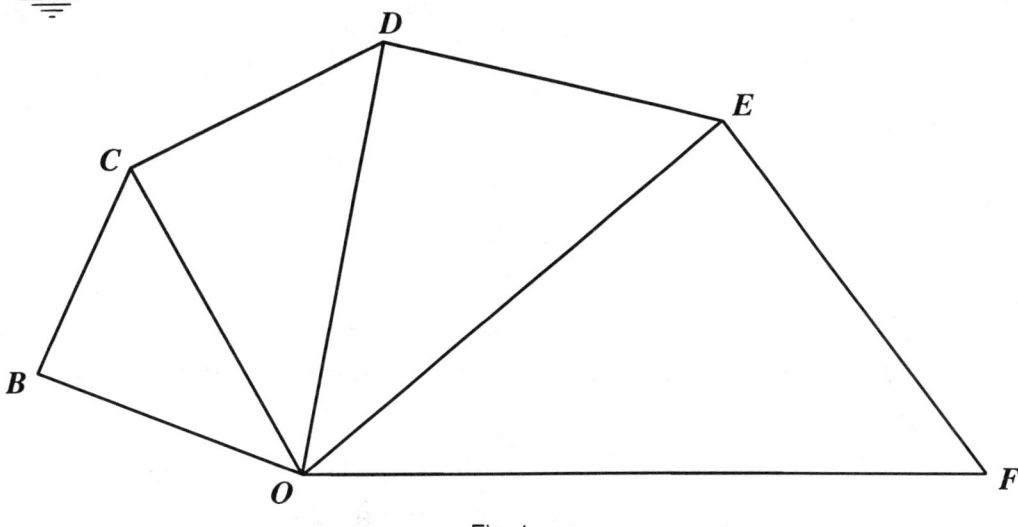

Fig. 1

1. In each column, list the congruent pairs of segments and the congruent pairs of angles in the two triangles by comparing the cut-out triangles one pair at a time. Note the example.

   △OBC and △OCD   △OCD and △ODE   △ODE and △OEF

   $\overline{OB} \cong \overline{CD}$

   $\overline{OC} \cong \overline{OC}$

   $\angle CBO \cong \angle DCO$

   $\angle BCO \cong \angle CDO$

   $\angle BOC \cong \angle COD$

2. What patterns can you find in the relationships between the sides and the angles of the consecutive triangles in the spiral? _____

3. If a △OFG is to be connected to △OEF to extend the spiral, what relationships should exist between △OFG and △OEF? _____

4. If a △OAB is to be connected to △OBC to extend the spiral inward, what relationships should exist between △OAB and △OBC? _____

The editors wish to thank Maurice Burke, Department of Mathematical Sciences, Montana State University, Bozeman, MT 59715, for writing this issue of *NCTM Student Math Notes*.

5-con Triangles—Continued

5. Using your cut-out triangles to help you with any necessary measurements, construct △*OFG* and △*OAB*.

6. △*OBC* and △*OCD* are not congruent. However, △*OBC* and △*OCD* are called a *5-con pair of triangles,* since exactly five of their six parts (two sides and three angles) are congruent. Name the other 5-con pairs of triangles in figure 1. _____
_____

Let's find out how to construct other spirals consisting of 5-con triangle pairs.

7. In each of the following sequences, find a pattern and use it to determine the next three numbers in the sequence.

   a) 6, 18, 54, 162, ____, ____, ____, . . .
   What pattern did you find? _____

   b) 16, 24, 36, 54, ____, ____, ____, . . .
   What pattern did you find? _____

8. Attempt to construct triangles whose three sides have the indicated lengths in millimeters. (A compass would be helpful.)

   a) 6, 18, 54   b) 18, 54, 162   c) 16, 24, 36   d) 24, 36, 54

5-con Triangles—Continued

9. Why is it impossible to construct triangles in cases 8a and 8b? _____
   _____

10. Cut out and compare the angles of triangles formed in cases 8c and 8d. What can you conclude about the triangles?
    _____
    _____

11. 16, 24, 36, and 54 are the first four terms of the sequence in 7b. Use additional terms from this sequence to guide you in constructing two more 5-con triangles on a separate sheet of paper. What are the lengths of their sides?
    _____

12. The sequences in 7a and 7b are both *geometric sequences,* since their numbers follow the pattern a, ar, ar^2, ar^3, ..., with $a > 0$. They are increasing sequences, since $r > 1$. Fill in the blanks:

    In the sequence in 7a, $a = $ _____, $r = $ _____.
    In the sequence in 7b, $a = $ _____, $r = $ _____.

    Some increasing geometric sequences like that in 7b generate spirals of 5-con triangles, and some like that in 7a do not. The generation of a spiral depends on whether three consecutive numbers in the sequence satisfy the triangle inequality, that is, that the sum of the lengths of any two triangles is greater than the third. The following theorem reveals the conditions under which a 5-con-triangle spiral can be formed.

    > **5-CON THEOREM:** Two triangles will form a 5-con pair if and only if the measures of their sides can be expressed as (a, ar, ar^2) and (ar, ar^2, ar^3) where $a > 0$ and $1 < r < (1 + \sqrt{5})/2$. Furthermore, an increasing geometric sequence $a, ar, ar^2, ar^3, \ldots,$ will generate a 5-con spiral of triangles if and only if $a > 0$ and $1 < r (1 + \sqrt{5})/2$.

13. Measure the sides of $\triangle OBC$ and $\triangle OCD$ in millimeters to see if the measurements satisfy the 5-con theorem. (Remember, your measurements will be only approximate.)

    $\triangle OBC$: _____, _____, _____ $\triangle OCD$: _____, _____, _____

14. Suppose $\triangle KLM$ has sides that measure 27, 36, and 48. Suppose $\triangle NPQ$ has sides of lengths 36, 48, and 64. Do these triangles form a 5-con pair? Justify your answer. _____
    _____

15. Is it possible to construct two triangles that have exactly three sides and two angles congruent? Why or why not?
    _____
    _____

16. Is it possible to construct two isosceles triangles that form a 5-con pair of triangles? Why or why not? _____
    _____

Teaching with *Student Math Notes:* Volume 2

Can you...

- construct a pair of 5-con right triangles?
- construct a pair of 5-con acute triangles?
- use the Geometric Supposer to study 5-con triangles?
- explain how △*OBC* and △*OCD* are not congruent and yet △*OBC* can have two sides and their included angle congruent to two sides and an angle of △*OCD* and still not contradict the SAS property?
- explain how △*OCD* and △*OBC* are not congruent and yet △*OCD* can have two angles and their included side congruent to two angles and a side of △*OBC* and still not contradict the ASA property?
- prove that if $0 < a < ar < ar^2$ and if the lengths $a, ar,$ and ar^2 satisfy the triangle inequality, then $1 < r < (1 + \sqrt{5})/2$?
- prove the 5-con theorem?

Did you know that...

- the ideas of 5-con triangles can be extended to 7-con quadrilaterals and 9-con pentagons, and so on?
- $(1 + \sqrt{5})/2$ is the golden ratio?
- when r equals the square root of the golden ratio the 5-con triangles formed by the sequence $a, ar, ar^2, ar^3, \ldots$, are right triangles?
- you can read all about 5-con triangles in the *Mathematics Teacher*? See the References.

References

- Pawley, Richard. "5-con Triangles." *Mathematics Teacher* 60 (May 1967):438–42.
- Pagni, David, and Gerald Gammon. "The Golden Mean and an Intriguing Congruence Problem." *Mathematics Teacher* 74 (December 1981):725–28.

Teacher Notes

Page 107. Some students may not see the spiral pattern in the figure. Be sure to point it out before they begin their cutting. The piece of the spiral shown here passes through points B, C, D, E, and F but not through O. As shown, it is actually a polygonal spiral formed from connected line segments.

Page 108. Have students cut out triangles OAB and OFG and use them with those cut from the previous page to form the polygonal spiral $ABCDEFG$. The complete spiral continues turning counterclockwise inward around point O with shorter and shorter segments and also clockwise outward with longer and longer segments.

Page 109. After completing question 11, have students cut out these triangles and those from question 8 and arrange the six of them into another polygonal spiral. Again, emphasize that the common vertex is not part of the spiral and that smaller and smaller and larger and larger triangles can always be added to the set. As the number of triangles increases, the spiral will begin to bend back around itself, and at that point, triangles will begin to overlap each other.

Page 112. This activity offers a good review of algebraic manipulation and computation with radicals, including finding the sum of an infinite, convergent geometric sequence where the ratio of successive terms contains a radical.

Answers

Page 107.
1. $\overline{OC} \cong \overline{DE}, \overline{OD} \cong \overline{OD}, \angle OCD \cong \angle ODE, \angle CDO \cong \angle DEO, \angle COD \cong \angle DOE$;
   $\overline{OD} \cong \overline{EF}, \overline{OE} \cong \overline{OE}, \angle ODE \cong \angle OEF, \angle DEO \cong \angle EFO, \angle DOE \cong \angle EOF$.

2. One pair of sides in common, three pairs of angles congruent, a second pair of sides congruent

3. $\angle OFG \cong \angle OEF, \overline{FG} \cong \overline{OE}, \overline{OF} \cong \overline{OF}, \angle FGO \cong \angle EFO, \angle FOG \cong \angle EOF$.
4. $\angle OAB \cong \angle OBC, \overline{OA} \cong \overline{BC}, \overline{OB} \cong \overline{OB}, \angle ABO \cong \angle BCO, \angle AOB \cong \angle BOC$.

Page 108.
5. In $\triangle OAB$, $AO = BC$, $BO = BO$, and $AB \approx 24$ mm. In $\triangle OFG$, $OF = OF$, $FG = OE$, and $OG \approx 104.5$ mm.
6. $\triangle OCD$ and $\triangle ODE$; $\triangle ODE$ and $\triangle OEF$
7. a) 486, 1458, 4374. Each term is three times the previous term.
   b) 81, 121.5, 182.25. Each term is 1.5 times the previous term.

Page 109.
9. The sum of the lengths of two sides of the triangle is less than the third side.
10. They are 5-con triangles; three pairs of angles are congruent, and two pairs of sides are congruent.
11. 36, 54, 81 and 54, 81, 121.5
12. 6, 3; 16, 3/2
13. $\triangle OBC$: 30, 37.5, 46.9; $\triangle OCD$: 37.5, 46.9, 58.6
14. 27, 36, 48, 64, $a = 27$, and $r = 4/3$. Yes, they form a 5-con pair—the 5-con theorem is satisfied.
15. No, because the sum of the measures of the three angles of a triangle is 180 degrees, and if two pairs of the angles were congruent, the third pair would also have to be congruent.
16. No, because if the triangle was isosceles then in the triple a, ar, ar^2, the r would equal 1; thus, the two triangles would be congruent and equilateral.

Page 112 (Extension).
1. 2; $\sqrt{5}-1$; $3-\sqrt{5}$; $2\sqrt{5}-4$; $7-3\sqrt{5}$

2. $(\sqrt{5}-1)/2$; $(3-\sqrt{5})/2$

3. The geometric series converges. $r = (3-\sqrt{5})/2$ $\qquad S = \dfrac{4}{1-\dfrac{3-\sqrt{5}}{2}} = 2 + 2\sqrt{5}$

   $A = lw = (1+\sqrt{5}) \times 2 = 2 + 2\sqrt{5}$

4. $\pi, \dfrac{\sqrt{5}-1}{2}\pi, \dfrac{3-\sqrt{5}}{2}\pi, (\sqrt{5}-2)\pi, \dfrac{7-3\sqrt{5}}{2}\pi$

5. $r = \dfrac{\sqrt{5}-1}{2}$ converging geometric series

   $S = \dfrac{\pi}{1-\dfrac{\sqrt{5}-1}{2}} = \dfrac{3+\sqrt{5}}{2}\pi$

6. $\dfrac{9+4\sqrt{5}}{8}$ to the right and $\dfrac{2+\sqrt{5}}{8}$ up

Extension—GOLDEN SPIRAL

A special set of squares can be placed in a certain way to form a golden spiral. Begin with a rectangle with sides in the golden ratio. Draw in the squares as shown. Then connect pairs of opposite vertices by drawing 90° arcs from a third vertex of each square.

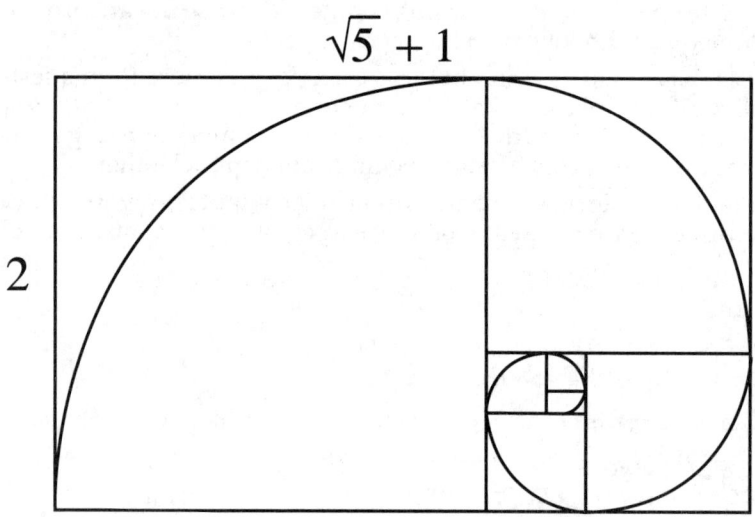

1. Find, algebraically, the length of a side of each of the six squares shown in the figure.

2. In each case, what is the ratio of a side of one square to that of the next smaller square? What is the ratio of areas of successive squares?

3. Use a converging geometric series to show that the sum of the areas of all such squares that can be formed is exactly equal to that of the original rectangle.

4. Find the length of each of the arcs shown in the first six squares drawn in the figure.

5. What would be the total length of the complete spiral if the process were continued ad infinitum?

6. About what specific point is the golden spiral turning? Give its location in relation to the initial point on the lower left of the original rectangle.

Striking Sequences

Throughout the ages, people have been fascinated with numbers in sequence. The *striking sequences* are formed from other sequences of numbers by striking out some elements and performing operations on the remaining sequences. The striking sequences given here were first discussed by the German number theorist Alfred Moessner in 1951 and have been extended and explained by the American number theorist Calvin Long, Washington State University. Try your hand at the following striking sequences.

I. Use the natural numbers and strike out every second number starting with 2.

1 2 3 4 5 6 7 8 9 10 11 12 13 14 15 16 17 . . .

Use the remaining numbers to make a sequence of partial sums from the striking sequence. The first four terms are given here:

$$1 = 1$$
$$1 + 3 = 4$$
$$1 + 3 + 5 = 9$$
$$1 + 3 + 5 + 7 = 16$$

The terms found by determining the partial sums as illustrated form a new sequence.

1. Use the terms you found to complete writing the first nine terms of this new sequence.

   1, 4, 9, 16, _____ , _____ , _____ , _____ , _____ , . . .

2. Describe the apparent pattern in the sequence that consists of the partial sums.

   _____

   _____

II. Construct another striking sequence from the natural numbers by striking out every third number starting with 3.

1 2 3 4 5 6 7 8 9 10 11 12 13 14 15 16 17 . . .

As in part I, form partial sums from this striking sequence. The first three terms are $1 = 1$, $1 + 2 = 3$, and $1 + 2 + 4 = 7$.

3. List the first ten terms in the sequence of partial sums.

   1, 3, 7, _____ , _____ , _____ , _____ , _____ , _____ , _____ , . . .

The editors wish to thank Carole B. Lacampagne, Department of Mathematical Science, Northern Illinois University, DeKalb, IL 60115, for writing this issue of *NCTM Student Math Notes*.

4. Next construct a new striking sequence from the partial-sums sequence in step 3 by striking out every other number starting with the second term.

5. Form a new sequence of partial sums from the sequence obtained in step 4.

6. Describe the apparent pattern in the sequence obtained in step 5.

III. Consider a striking sequence formed from the natural numbers 1–24 using all steps a through f:
   (a) Strike out every fourth number.
   (b) Make the sequence of partial sums.
   (c) Strike out every third number from the sequence formed in step b.
   (d) Make the sequence of partial sums from the sequence formed in step c.
   (e) Strike out every other number from the sequence formed in step d.
   (f) Make the sequence of partial sums from the sequence formed in step e.

7. Predict what the final sequence will be.

8. Construct the striking sequence using all the steps a through f in the order given.

9. What would be the result of repeating the process in part III by striking out every fifth number beginning with 5. (You may wish to start with a longer sequence of natural numbers.) Construct the sequence.

IV. In parts I–III striking sequences were obtained from the sequence of natural numbers. Next use the sequence of odd counting numbers and strike out every other number starting with 3.

1 2 3 4 5 6 7 8 9 10 11 12 13 14 15 16 17 . . .

10. Form the sequence of partial sums from the striking sequence of odd numbers obtained above.

The sequence of partial sums obtained in step 10 may not look familiar. However, consider factoring each term as follows:

$$1, \quad 2 \cdot 3, \quad 3 \cdot 5, \quad 4 \cdot 7, \ldots$$

11. Predict the factorization of the next six terms in the sequence.

   ———, ———, ———, ———, ———, ———

12. Verify your answer by extending the sequence in step 10.

V. Next construct a striking sequence using the odd numbers, striking every third number, getting partial sums, striking every other number of the partial-sums sequence, and getting a new sequence of partial sums.

13. What is the final sequence you obtained?

14. What is the pattern found by factoring the sequence?

VI. Next let's construct a striking sequence using the natural numbers. Strike out the first number, then the second number after that, then the third number after that, the fourth number after that, and so on.

15. List the sequence obtained.

16. Make a sequence of partial sums from the sequence obtained in step 15.

17. Construct a new striking sequence from the one in step 16 in the same fashion as before: Strike out the first number, then the second number after that, then the third number after that, and so on; then take the partial sums to form another sequence and write it.

18. Repeat the process on the sequence in step 17. Show the sequence of partial sums.

19. Repeat the process on the sequence in step 18. Show the sequence of partial sums.

20. Form a new sequence starting with 1, made up of the *first* stricken entries of the sequences in steps 16, 17, 18, and 19. Continue the pattern.

    1,   2,   6, _____

21. Use the factorizations of the first three terms of the sequence in step 19 to try to discover a pattern. Then factor the next two terms similarly.

    1,   1 · 2,   1 · 2 · 3, _____

22. Describe the sequence obtained in step 20.

Did you know that . . .

- a sequence is said to be *arithmetic* if the difference between consecutive terms is the same? The natural numbers form an arithmetic sequence because the common difference between terms is 1. The odd numbers form an arithmetic sequence because the common difference between terms is 2. Any sequence of the form $a, a + d, a + 2d, a + 3d, \ldots, a + nd$ is arithmetic. The first term is a, and the common difference between terms is d.

Teaching with *Student Math Notes:* Volume 2

- the sum of the first n terms of an arithmetic sequence is
$$S_n = (1/2)n[2a + (n - 1)d]?$$
- as a school boy, Karl Freidrich Gauss discovered the sum of the first n natural numbers?

Can you . . .

- generalize the results in parts IV and V? What will the final sequence be if you start with the odd counting numbers, strike out every kth number, get partial sums, strike out every $(k - 1)$st number, get partial sums, and so on?
- find Gauss's formula and use it to prove the formula for the sum of the first n terms of an arithmetic sequence?
- write an arithmetic sequence of your choice, construct a striking sequence from it by striking out every kth number, get partial sums, and so on? Can you generalize about your striking sequence?
- write an arithmetic sequence and construct a striking sequence, as in part IV?
- write a computer program to complete the process in part VI? (Since the numbers in your sequence will grow rapidly, you may need to use double-precision variables.)

Have you read . . . ?

- Long, Calvin. "Mathematical Excitement—the Most Effective Motivation." *Mathematics Teacher* 75 (May 1982):413–15.
- ———. "On the Moessner Theorem on Integral Powers." *American Mathematical Monthly* 74 (October 1966):846–51.
- ———. "Strike It Out—Add It Up." *Mathematical Gazette* 66 (December 1982):273–77.
- Moessner, Alfred. "Eine Bemerkung uber die Potenzen der naturlichen Zahlen" [One Observation on the Power of Natural Numbers]. Munich: Bavarian Academy of Sciences and Mathematics. Nat. K1 S. -B., 29.

Teacher Notes

Page 113. One way to present these striking sequences is to place the natural numbers on file cards and arrange the set sequentially on the tray of the chalkboard. Then simply remove different subsets of cards, leaving the desired striking sequences to observe and study with the eye as well as the mind.

As a variation to Part II, form a new sequence by adding the successive pairs that remain after every third counting number starting with 3 is removed.

1 + 2	4 + 5	7 + 8	10 + 11	13 + 14	16 + 17	19 + 20	...
3	9	15	21	27	33	49	

Then form partial sums from this sequence.

$$3 = 3 \qquad 3 = 3 \times 1 \qquad = 3 \times 1^2$$
$$3 + 9 = 12 \qquad 12 = 3 \times 4 \qquad = 3 \times 2^2$$
$$3 + 9 + 15 = 27 \qquad 27 = 3 \times 9 \qquad = 3 \times 3^2$$
$$3 + 9 + 15 + 21 = 48 \qquad 48 = 3 \times 16 \qquad = 3 \times 4^2$$

Page 114. The process in Part IV can be simplified by initially showing only the odd counting numbers. Be certain students understand that it is not the striking sequence, such as 1, 5, 9, 13, 17, ... in step 10, but the corresponding sequence of partial sums that is of interest.

$$1 \qquad 1 + 5 = 6 \qquad 1 + 5 + 9 = 15 \qquad 1 + 5 + 9 + 13 = 28 \qquad ...$$

Page 115. Students gain valuable practice in following directions when working through the different activities on this page. Encourage individual explorations as well, where students start with their own arithmetic sequences and striking sequences and look for patterns in the corresponding partial sums.

Page 118. The same types of observations that were made from number sequences and patterns can be applied to geometric arrays and patterns as well. In some cases, students may need to draw pictures or build models for the various subsets to facilitate their counting.

Answers

Page 113.
1. 1, 4, 9, 16, 25, 36, 49, 64, 81
2. The sequence consists of the squares of the counting numbers: $1^2, 2^2, 3^2, 4^2, 5^2, 6^2, 7^2, 8^2, 9^2, ...$
3. 1, 3, 7, 12, 19, 27, 37, 48, 61, 75, ...

Page 114.
4. 1, 7, 19, 37, 61, ...
5. 1, 8, 27, 64, 125, ...
6. The sequence consists of the cubes of the counting numbers: $1^3, 2^3, 3^3, 4^3, 5^3, 6^3, 7^3, 8^3, ...$
7, 8. The sequence consists of the fourth powers of the counting numbers: 1, 16, 81, 256, 625, 1296, ...
9. The sequence consists of the fifth powers of the counting numbers: 1, 32, 243, 1024, 3125, ...
10. 1, 6, 15, 28, 45, 66, 91, ...
11. $5 \cdot 9, 6 \cdot 11, 7 \cdot 13, 8 \cdot 15, 9 \cdot 17, 10 \cdot 19$
12. 120, 153, 190, 231, ...

Page 115.
13. 1, 12, 45, 112, 225
14. $1, 2^2 \cdot 3, 3^2 \cdot 5, 4^2 \cdot 7, 5^2 \cdot 9, ...$
15. 2, 4, 5, 7, 8, 9, 11, 12, 13, 14, 16, 17, 18, 19, 20, 22, ...
16. 2, 6, 11, 18, 26, 35, 46, 58, 71, 85, 101, 118, 136, 155, 175, ...
17. 6, 24, 50, 96, 154, 225, 326, 444, 580, 735, ...
18. 24, 120, 274, 600, 1044, 1624, ...
19. 120, 720, 1764, ...
20. 1, 2, 6, 24, 120, ...
21. $1 \cdot 2 \cdot 3 \cdot 4, 1 \cdot 2 \cdot 3 \cdot 4 \cdot 5, ...$
22. The sequence consists of the factorials of the counting numbers: 1!, 2!, 3!, 4!, 5!, 6!, 7!, 8!, ...

Page 118 (Extension).

1. 1, 4, 9, 16, 25, 36
   These numbers are the perfect squares.
2. 1, 3, 6, 10, 15, 21
   These numbers are the triangular numbers.
3. 0:1; 1:3; 3:6; 6:10; 10:15; 15:21
   The ratios increase toward 1.
4. 1, 3, 6, 9, 12, 15
   Continue by adding 3 each time.
5. 1, 3, 7, 13, 21, 31
   These numbers come by adding triangular numbers as shown here:
   $$1 = 1$$
   $$3 = 3$$
   $$6 + 1 = 7$$
   $$10 + 3 = 13$$
   $$15 + 6 = 21$$
   $$21 + 10 = 31$$
6. 0:1; 1:3; 3:6; 7:9; 13:12; 21:15
   The ratio becomes large without bound.

Extension—STRIKING OUT

Triangular pieces can be used to build other triangles as shown here.

1. Count the small triangular pieces in each figure. What pattern do you see in the numbers that you get?

2. Strike out all the small triangular pieces that point down. Then count the small triangular pieces that remain in each figure. What pattern do you see in the numbers that you get?

3. What is the ratio of the number of pieces removed to those that remain in each figure shown?

Start again with the original figures shown.

4. Strike out all the border triangles. Count the remaining pieces in each figure. What pattern do you see in the numbers that you get?

5. This time strike out all but the border triangles. Count the remaining pieces in each figure. What pattern do you see in the numbers that you get?

6. What is the ratio of the number of interior triangles to the number of border ones in the figures shown? If the pattern were continued, how would this ratio change?

Invariants

The knobs of appliances, the heads of screws, and a road sign are seen in figure 1.

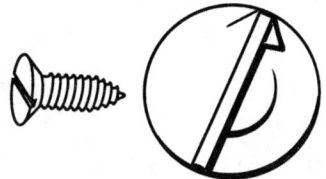  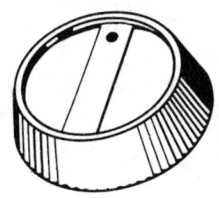 

Fig. 1

A two-dimensional representation of the tops of these common items can be depicted as a rectangle inscribed in a circle. The vertices of the rectangle are always on the circle, as in figure 2.

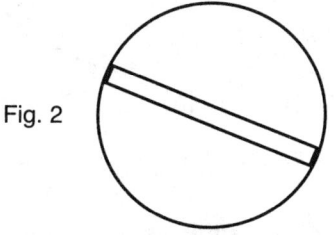

Fig. 2

Carefully inscribe five rectangles in the circle in figure 3 and draw one diagonal in each rectangle.

1. What common properties do all your rectangles in figure 3 share?

_____
_____
_____
_____
_____
_____
_____
_____
_____
_____
_____

Fig. 3

The editors wish to thank Melfried Olson, Western Illinois University, Macomb, IL 61455, and Glenn Bruckhart, Araphoe High School, Littleton, CO 80120, for writing this issue of *NCTM Student Math Notes*.

Other Invariants of Inscribed Quadrilaterals

All the properties found in question 1 are called *invariant properties* of this set of rectangles. One such property is that for any circle or circles having the same radii, the diagonals of all inscribed rectangles are the diameters of the circles and have the same length.

In figure 4 carefully inscribe one nonrectangular quadrilateral in each circle.

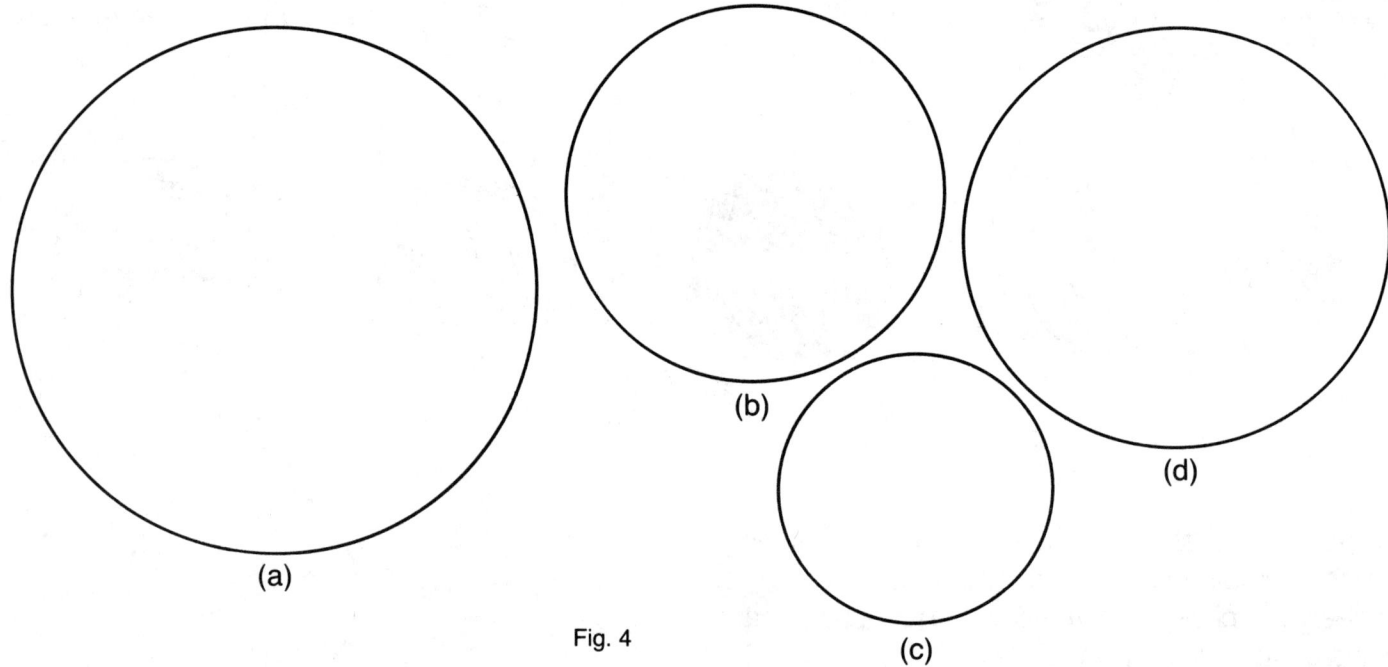

Fig. 4

2. Label each quadrilateral in figure 4 as *ABCD*. Draw diagonals $\overline{AC}$ and $\overline{BD}$ and let *E* be the intersection of diagonals $\overline{AC}$ and $\overline{BD}$. Record the measures of each of the following angles and segments in table 1 (measure segments in millimeters).

Table 1

	Figure 4a	Figure 4b	Figure 4c	Figure 4d
a) $m\angle ABC$				
b) $m\angle BCD$				
c) $m\angle CDA$				
d) $m\angle DAB$				
e) $m\angle DAB + m\angle BCD$				
f) $m\angle ABC + m\angle CDA$				
g) AE				
h) EC				
i) BE				
j) ED				
k) $AE \cdot EC$				
l) $BE \cdot ED$				

More Invariants

3. Use 2e and 2f to describe the sums of the measures of the opposite angles of each quadrilateral in table 1. What did you discover?

_____
_____
_____
_____

4. Use 2k and 2l to find the products of the lengths of the segments of the chords in table 1. What did you discover?

_____
_____
_____
_____

Table 2 shows several pairs of numbers whose product is 12.

Table 2

a	1	1.5	2	2.5	3
b	12	8	6	4.8	4

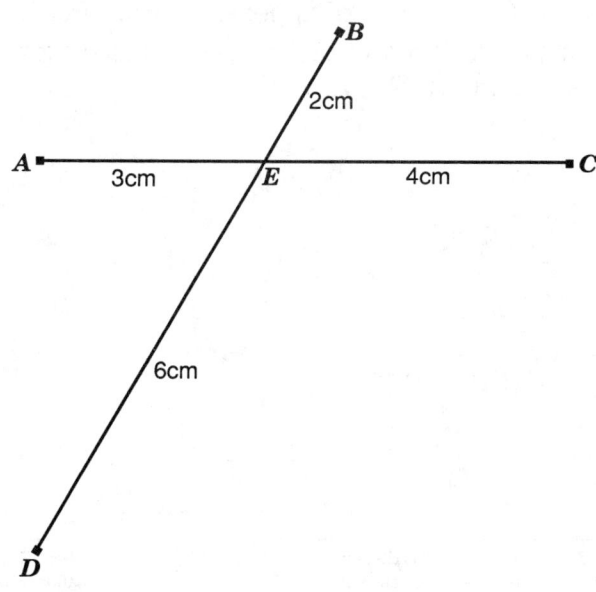

Fig. 5

In table 2, we have an invariant property that $a \cdot b = 12$ for each pair in a column. Can we relate this invariant property to circumscribed circles? In figure 5, two segments intersecting at E have been drawn. One segment, $\overline{AC}$, is 7 cm long and broken into pieces of lengths 3 cm and 4 cm. The other segment, $\overline{BD}$, is 8 cm long and broken into pieces of lengths 2 cm and 6 cm. (Note that the two pairs, 3 and 4 and 2 and 6, are from table 2 and that $3 \times 4 = 2 \times 6$.) Construct a circle that contains points A, B, C, and D. (Hint: Find the intersection of the perpendicular bisectors of $\overline{AC}$ and $\overline{BD}$.)

Use another two pairs, 1.5 and 8 and 3 and 4, to draw segment $\overline{AC}$ of length 9.5 cm, or (1.5 + 8), and segment $\overline{BD}$ of length 7 cm, or (3 + 4), which intersect at point E as in figure 5. Construct a circle that contains points A, B, C, and D.

Teaching with *Student Math Notes:* Volume 2

5. Will the circle you obtain be congruent to the circle a friend makes? _____

   Repeat this experiment using other pairs from the table.

6. Write an invariant property for intersecting segments (the products of whose parts are the same).

   _____

   _____

   _____

Did You Know That . . .

- defining properties are invariant properties, but not all invariant properties are defining properties of geometric objects?
- transformational geometry was begun as a study of invariant properties?
- the Pythagorean theorem is an invariant property of right triangles?
- in an inscribed quadilateral, the angle between a side and a diagonal is equal to the angle between the opposite side and the other diagonal?
- in an inscribed quadilateral, the product of the diagonals is equal to the sum of the products of the opposite sides?

Can You . . .

- find other invariant properties involving segments and circles?
- find other invariant properties involving measures of angles inscribed in circles?
- use the Geometric preSupposer software Points and Lines to do the explorations suggested in figure 5?
- describe the locus of the centers of the circles you constructed in question 5 if $\overline{BD}$ is rotated at E until it is collinear with $\overline{AC}$?

NCTM STUDENT MATH NOTES is published as a supplement to the **NEWS BULLETIN** by the National Council of Teachers of Mathematics, 1906 Association Drive, Reston, VA 22091. The five issues a year appear in September, November, January, March, and May. Pages may be reproduced for classroom use without permission.

Editor:	**Johnny W. Lott,** University of Montana, Missoula, MT 59812
Editorial Panel:	**Ena Gross,** Virginia Commonwealth University, Richmond, VA 23284
	Judy Olson, Western Illinois University, Macomb, IL 61455
	Daniel J. Teague, North Carolina School of Science and Mathematics, Durham, NC 27705
Editorial Coordinator:	**Joan Armistead**
Production Assistants:	**Ann M. Butterfield, Sheila C. Gorg**

Printed in U.S.A.

Teacher Notes

Page 119. At the same time that you focus on the invariant properties that students list for question 1, contrast the properties with some of the variation that does take place in going from one rectangle to another. For example, the diagonal of every rectangle drawn in the circle has the same length, that of the diameter of the circle, but as the width of the rectangle increases, the length decreases and both the perimeter and the area change.

Page 120. Encourage careful accurate measuring with ruler and protractor. Even though the measurements are approximate, students may recognize invariant relationships among the angles in the table. However, the invariant relationship among the parts of the chords will likely not be apparent until attention is focused on their products on the next page.

Page 121. A circle can be drawn through any three noncollinear points in a plane. Given four noncollinear points, the products of the two segments on each diagonal must be equal for the four points to lie on a circle. A good test of the students' understanding of these two ideas is to give them three noncollinear points and ask where a fourth point must be so that the two products of the pairs of segments are equal. Clearly, any fourth point on the circle located by the other three points suffices.

Page 122. In comparing results for question 5, students will recognize that the radii of the circles drawn from the segments for the two given chords depend on the angles formed by these chords. As a pair of vertical angles approaches 180 degrees each, the corresponding radii of the circle become large without bound. This becomes apparent when one realizes that the center of the circle must lie on the intersection of the perpendicular bisectors of the two segments, but these perpendicular bisectors are approaching parallelism.

Answers

Page 119. 1. Answers may vary. Examples may include the following: All diagonals are the same length; all angles are right angles; two pairs of congruent chords and arcs are determined by each rectangle; the opposite sides are equal, the opposite sides are parallel.

Page 120. 2. Answers will vary.

Page 121. 3. Opposite angles are supplementary; that is, the sum of opposite angles is 180 degrees.

4. The product of the lengths of the segments of one diagonal equals the product of the lengths of the segments of the other diagonal. From the actual measurements used, more than likely the products will be close, not exactly equal.

Page 122. 5. Probably not

6. If two segments AB and CD intersect at a single point E that is not the endpoint of either segment, such that $a_1 \times b_1 = c_1 \times d_1$ (in figure), then a circle can be drawn that contains the endpoints of both segments. However, the radius of the circle is dependent on the angle "between" the segments.